Dynamic Time Series Models using R-INLA

Dynamic Time Series Models using R-INLA: An Applied Perspective is the outcome of a joint effort to systematically describe the use of R-INLA for analysing time series and showcasing the code and description by several examples. This book introduces the underpinnings of R-INLA and the tools needed for modelling different types of time series using an approximate Bayesian framework.

The book is an ideal reference for statisticians and scientists who work with time series data. It provides an excellent resource for teaching a course on Bayesian analysis using state space models for time series.

Key Features:

- Introduction and overview of R-INLA for time series analysis.
- Gaussian and non-Gaussian state space models for time series.
- State space models for time series with exogenous predictors.
- Hierarchical models for a potentially large set of time series.
- Dynamic modelling of stochastic volatility and spatio-temporal dependence.

Nalini Ravishanker is a professor in the Department of Statistics at the University of Connecticut, Storrs, USA.

Balaji Raman is a statistician at Cogitaas AVA, Mumbai, India.

Refik Soyer is a professor in the Department of Decision Sciences at The George Washington University, Washington D.C., USA.

Dynamic Time Series Models using R-INLA
An Applied Perspective

Nalini Ravishanker, Balaji Raman and Refik Soyer

CRC Press
Taylor & Francis Group
Boca Raton London New York

CRC Press is an imprint of the
Taylor & Francis Group, an **informa** business

A CHAPMAN & HALL BOOK

First edition published 2023
by CRC Press
6000 Broken Sound Parkway NW, Suite 300, Boca Raton, FL 33487-2742

and by CRC Press
4 Park Square, Milton Park, Abingdon, Oxon, OX14 4RN

CRC Press is an imprint of Taylor & Francis Group, LLC

Library of Congress Cataloging-in-Publication Data
Names: Ravishanker, Nalini, author. | Raman, Balaji, author. | Soyer, Refik, author.
Title: Dynamic time series models using R-INLA : an applied perspective / Nalini Ravishanker, Balaji Raman and Refik Soyer.
Description: First edition. | Boca Raton : CRC Press, 2022. | Includes bibliographical references and index.
Identifiers: LCCN 2022004440 (print) | LCCN 2022004441 (ebook) | ISBN 9780367654276 (hardback) | ISBN 9780367680626 (paperback) | ISBN 9781003134039 (ebook)
Subjects: LCSH: Time-series analysis. | Bayesian statistical decision theory. | R (Computer program language) | Laplace transformation. | Mathematical models.
Classification: LCC QA280 .R38 2022 (print) | LCC QA280 (ebook) | DDC 519.5/502855133--dc23/eng20220421
LC record available at https://lccn.loc.gov/2022004440
LC ebook record available at https://lccn.loc.gov/2022004441

ISBN: 978-0-367-65427-6 (hbk)
ISBN: 978-0-367-68062-6 (pbk)
ISBN: 978-1-003-13403-9 (ebk)

DOI: 10.1201/9781003134039

Typeset in LM Roman
by KnowledgeWorks Global Ltd.

Publisher's note: This book has been prepared from camera-ready copy provided by the authors.

To

My family and friends (NR)

My teachers (BR)

My family (RS)

Contents

Preface

The Integrated Nested Laplace Approximation (INLA) is a well-known method for carrying out approximate Bayesian inference. Through the `R-INLA` package and its well maintained companion website, INLA has established itself as an alternative to other methods for Bayesian analysis, such as Markov chain Monte Carlo (MCMC) or Variational Bayes, primarily because of its speed and ease of use. INLA is a fast alternative to MCMC methods for Bayesian modeling in complex problems when exact, closed form solutions are not available - that is most of the data analysis in practice today!

This book has grown out of our experience over several years and is aimed at applied scientists or applied statisticians who work with time series data in various forms, sizes, and shapes, and would like to get fast and accurate, though approximate, solutions. The goal is to provide a guide for using the `R-INLA` package to analyze different classes of time series in a dynamic Bayesian framework.

The INLA methodology was introduced in Rue et al. (2009) for models that can be expressed as latent Gaussian Markov random fields (GMRFs). Although this may seem restrictive, it is amazing how many families of models that are used in practice for large, complex data can be handled in this framework. Dynamic models for time series is one domain where `R-INLA` can be extremely effective.

Why read this book?

Dynamic modeling of data observed over time is increasingly important as data with temporal dependence is frequently available in various disciplines. This book describes the use of an approximate Bayesian framework using `R-INLA` for dynamic modeling of time series under various situations. The details provided in different chapters should be useful for practitioners who are interested in using `R-INLA` for modeling.

Structure of the book

Chapter 1 introduces the Bayesian framework and the dynamic Bayesian framework for time series, while Chapter 2 describes the Integrated Nested Laplace Approximation (INLA) framework for carrying out fast, approximate Bayesian modeling and forecasting. Chapter 3 sets up the `R-INLA` for various univariate dynamic linear models, while Chapter 4 illustrates use of these models via a univariate time series from software engineering, and explains prior specification, parameter, and hyperparameter estimation, model selection, and forecasting. A careful reading of Chapter 3 and Chapter 4 will enable a reader to learn how to set up many items in R-INLA and enhance them in later chapters. Chapter 5 shows how to fit structural time series models and include exogenous time series as predictors, both in the observation equation and in the state equation. Chapter 6 sets up hierarchical DLMs for panel time series. We describe many aspects of R-INLA which will be useful in later chapters. Chapters 7, 8, and 9 describe non-Gaussian state space models. While Chapter 7 describes models for continuous, positive-valued responses and for time series of proportions taking values in the interval $(0, 1)$, Chapters 8 and 9 respectively model binary/categorical time series of counts. Chapter 10 shows how to use R-INLA for stochastic volatility modeling with application to financial returns, although these methods can be used for any nonlinear growth modeling. Chapter 11 looks at data that occur over space and time, and builds models that handle spatial and temporal effects additively, as well as through various types of interactions. In Chapter 12, we model multivariate Gaussian DLMs by using the augmented structure for representing the responses and obtaining fitted values and forecasts through the functions inla.make.lincomb() and inla.posterior.samples(). We also show an example of using rgeneric() to build MV DLM for which a template does not currently exist in R-INLA. Chapter 13 illustrates how to extend the ideas in Chapter 6 to build hierarchical DLMs for panels of vector-valued time series of counts. In this chapter, we also show how we can build level correlated models for vector-valued counts, addressing a topic that is becoming increasingly important in various application domains. Chapter 14 pulls together some often used information that can serve as a quick look-up guide for users, while the appendices in some chapters provide some useful information that is tangential to the R-INLA modeling.

Software information and conventions

This is an R book (Team, 2018) and all codes are written in R version 4.1.0 (2021-05-18). The IDE used is RStudio (RStudio Team, 2021). All the code in this book are developed using INLA_21.02.23. The book is compiled using the **bookdown** package (Xie, 2016). We have used the following typographical conventions - package names are in typewriter font (e.g., **ggplot2**). Function names followed by parentheses are in typewriter font (e.g., inla()). Function arguments are also in typewriter font (e.g., control.compute). User specified variable or data names are in italics (e.g., *y*, *Musa*).

Plots have been constructed using plot(), ts.plot(), etc. from R or by using functions that we have created such as tsline() and multiline.plot(). In our functions we have customized ggplot(). These are available in Chapter 14 as well as in our GitHub link https://github.com/ramanbala/dynamic-time-series-models-R-INLA. The online version of this book is available at https://ramanbala.github.io/dynamic-time-series-models-R-INLA/.

Acknowledgments

We are very grateful to Håvard Rue and the R-INLA team for their expert and timely help with different aspects of the code and output. We are also thankful to Bradley Boehmke (https://GitHub.com/bradleyboehmke) for his generous help with our markdown struggles. Excellent books that describe R-INLA (by Virgilio Gomez-Rubio, Paula Moraga, and Marta

Blangiardo and Michela Cameletti) have been extremely useful to us. We are grateful to the authors. Several current and past students at the University of Connecticut have been involved in our R-INLA journey. A special note of thanks to Chiranjit Dutta, Shan Hu (posthumous), and Volodymyr Serhiyenko for codes related to multivariate models. We also thank Renjie Chen and Patrick Toman for reading earlier versions and making suggestions. We thank the organizers and participants of the following workshops - Bayesian Analysis of Time Series Data using R (ISBIS 2019, Kuala Lumpur, Malaysia) and Workshop on DLM using R (YBIS 2019, Istanbul, Turkey) for their feedback which motivated us to start working on this book. We thank Kamal Sen and Venu Gorti (Cogitaas AVA, Mumbai, India) for allowing us to use their computing facilities to run our codes. We are also grateful to the ITS staff at the University of Connecticut, Storrs, U.S.A. for their help in enabling us to run R-INLA code on the HPC cluster. We appreciate all the help we received from David Grubbs at Chapman & Hall/CRC. Finally, we thank our families for their constant support and encouragement - my family (NR); my mother, Shraddha and Pinakin for letting me do my R-INLA *homework* peacefully (BR); Ayşegül, Deniz, and Demir (RS).

Nalini Ravishanker, Balaji Raman, and Refik Soyer

1

Bayesian Analysis

1.1 Introduction

This chapter gives a brief introduction to Bayesian framework in statistical analysis. In the Bayesian approach, the posterior probability distributions of parameters (which are treated as random quantities) have a central role. Based on the likelihood function of the parameters, given data, and a prior distribution on the parameters, the posterior distribution is obtained via an application of Bayes' theorem. In many complex modeling situations, including most time series analyses, closed-form expressions for the posterior distributions are rarely available. One exception of course, is the Gaussian Dynamic Linear Model (DLM), for which Kalman filtering and smoothing equations offer closed form expressions for the posterior distribution of the unknown state variable. To handle complex situations, sampling based Bayesian approaches can be used for posterior inference and prediction, see Tanner and Wong (1987), Gelfand and Smith (1990), and Gamerman and Lopes (2006). General solutions via Markov Chain Monte Carlo (MCMC) methods have been provided to practitioners by software such as R, WinBUGS (Lunn et al., 2000), STAN,[1] or JAGS (Plummer et al., 2003). While MCMC is an asymptotically exact method for posterior analysis, it does not scale well and can be slow in targeting random samples from the joint posterior of the parameters, especially in high dimensions and complex data situations. The computational complexity can be high even under distributed (parallel) computing environments. Under limited sampling, convergence of the MCMC algorithms may become an issue, and one cannot have much confidence in the accuracy of the results.

An alternative to MCMC for large, complex problems is provided by *variational inference* (Blei et al., 2017). See Korobilis and Koop (2018) for a discussion of variational Bayesian (VB) inference for state-space models, and Berry and West (2020) for use of VB in modeling time series of multivariate counts. Application of VB methods may not be straightforward for many practitioners due to limited availability of software at the current time.

The Integrated Nested Laplace Approximation (INLA) (Rue et al., 2009) provides a fast alternative to MCMC methods for Bayesian modeling in complex problems when exact, closed form solutions are not available. Research on the use of INLA within R has been prolific in recent years, with several books and articles describing how to use R-INLA in different settings. In this book, we describe the use of R-INLA for state space modeling of time series data. The use of dynamic models for handling a variety of time series will be of interest to a wide audience of academics and practitioners who work with data that are observed over time. Ruiz-Cárdenas et al. (2012) describe the use of R-INLA for Dynamic Linear Models (DLMs).

---

[1]https://mc-stan.org/docs/2_29/reference-manual-2_29.pdf

Section 1.2 gives a brief review of the Bayesian framework. In Section 1.3, we review the setting for a Bayesian analysis of time series data. Section 1.4 introduces Gaussian DLMs, while Section 1.5 gives an overview of hierarchical Gaussian and non-Gaussian models.

1.2   Bayesian framework

As noted by Lindley (1983), the Bayesian approach is based on the premise that *"the only satisfactory description of uncertainty is by means of probability."* This suggests that uncertainty about any unknown quantity, observed or unobserved (such as parameters), should be described probabilistically, and uncertainties must be combined using the rules of probability.

To introduce some notation, we let Y be a random variable, i.e., the observable random quantity, with realization y, and Θ denote an unknown (latent) parameter. The unobserved random quantity Θ can be a scalar or a vector. Suppose that a probability model is specified with a p.d.f. (or p.m.f.) $p(y|\Theta)$. For example, if $Y \sim N(\mu, \sigma^2)$, then $\Theta = (\mu, \sigma^2)$ and if $Y \sim Poisson(\lambda)$, then $\Theta = \lambda$. In Chapter 1 – Appendix, we give a brief review of conditional distributions. Before observing any data, uncertainty about the unknown quantity Θ is described by $\pi(\Theta)$ which is referred to as the *prior distribution* of Θ as it reflects our *a priori* knowledge/belief about Θ. After we observe the data, $\boldsymbol{y}^n = (y_1, \dots, y_n)'$, we update our prior beliefs about Θ to the posterior distribution $\pi(\Theta|\boldsymbol{y}^n)$ using the calculus of probability, i.e., via Bayes' theorem as

$$\pi(\Theta|\boldsymbol{y}^n) \propto \pi(\Theta)L(\boldsymbol{y}^n|\Theta), \qquad (1.1)$$

where $L(\boldsymbol{y}^n|\Theta)$ is the *likelihood function*. The likelihood $L(\boldsymbol{y}^n|\Theta)$ is obtained by evaluating the joint p.d.f. of $(Y_1, \dots, Y_n)$ at the observed data $\boldsymbol{y}^n$ for different values of Θ. Thus, it is a function of Θ and can be interpreted as a scale of comparative support given by the observed data $\boldsymbol{y}^n$ to various possible values of Θ. To reflect this, we find it more convenient to write the likelihood function as $L(\Theta; \boldsymbol{y}^n)$ and to define the log likelihood by $\ell(\Theta; \boldsymbol{y}^n)$.

The *likelihood principle* states that the totality of information about Θ provided by $\boldsymbol{y}^n$ is in the likelihood function, and plays an important role in the Bayesian framework. As a result, all inference about Θ will be based on the posterior distribution which considers only the observed data $\boldsymbol{y}^n$, but all possible values of Θ. The constant of integration for (1.1) is given by

$$m(\boldsymbol{y}^n) = \int L(\Theta; \boldsymbol{y}^n)\pi(\Theta)d\Theta, \qquad (1.2)$$

which is referred to as the *marginal likelihood* since it does not depend on Θ. Computation of the marginal likelihood follows from the rules of probability since

$$p(\boldsymbol{y}^n) = \int p(\boldsymbol{y}^n|\Theta)\pi(\Theta)d\Theta, \qquad (1.3)$$

where $p(\boldsymbol{y}^n|\Theta) = \prod_{j=1}^{n} p(y_j|\Theta)$. To evaluate the marginal likelihood $m(\boldsymbol{y}^n)$, we replace $p(\boldsymbol{y}^n|\Theta)$ by $L(\Theta; \boldsymbol{y}^n)$ in (1.3), once data become available. Prior to observing $\boldsymbol{y}^n$, $p(\boldsymbol{y}^n)$ is referred to as the *prior predictive distribution* of $(Y_1, \dots, Y_n)$.

Once the posterior distribution $\pi(\Theta|\boldsymbol{y}^n)$ is available, the posterior predictive distribution of a future (new) observation $\boldsymbol{y}^*$ is given by

$$p(\boldsymbol{y}^*|\boldsymbol{y}^n) = \int p(\boldsymbol{y}^*|\Theta)\pi(\Theta|\boldsymbol{y}^n)d\Theta. \tag{1.4}$$

As noted by Ebrahimi et al. (2010), the concept of prediction, which is usually an afterthought in classical statistics, arises naturally in the Bayesian approach as a consequence of using rules of probability. The predictive distribution is an essential component of the Bayesian approach in time series analysis.

1.2.1 Bayesian model comparison

Marginal likelihood

We defined the marginal distribution of the data $\boldsymbol{y}^n$ in (1.2). This concept plays a useful role in Bayesian model comparison when we must select from a set of models, $M_1, \ldots, M_K$. Let $\ell(\Theta; \boldsymbol{y}^n, M_k)$ denote the log likelihood under model M_k of Θ given the data $\boldsymbol{y}^n$, and let $\pi(\Theta|M_k)$ denote the prior distribution of Θ under model M_k. Each model may have its own set of parameters. Then, the marginal likelihood of model M_k is

$$m(\boldsymbol{y}^n|M_k) = \int L(\Theta; \boldsymbol{y}^n, M_k)\pi(\Theta|M_k)d\Theta. \tag{1.5}$$

The marginal likelihood is useful in many Bayesian modeling steps, such as in the acceptance ratio computation in the Metropolis-Hastings algorithm, in Bayesian model selection and model averaging, etc. However, in most cases, $p(\boldsymbol{y}^n|M_k)$ cannot be obtained analytically. Many methods have been proposed including those by Tierney and Kadane (1986) (via the Laplace approximation), Newton and Raftery (1994) (harmonic mean estimator), Chib (1995) (via MCMC), Marin et al. (2012) (using approximate Bayesian computation, ABC), Jordan et al. (1998) (variational methods), or Rue et al. (2009) (INLA).

Bayes factors

Consider two candidate models M_1 and M_2 to describe the given data $\boldsymbol{y}^n$. Using posterior model probabilities $\pi(M_k|\boldsymbol{y}^n)$ for $k = 1, 2$, we can write the *posterior odds* as

$$\frac{\pi(M_1|\boldsymbol{y}^n)}{\pi(M_2|\boldsymbol{y}^n)} = \frac{m(\boldsymbol{y}^n; M_1)}{m(\boldsymbol{y}^n; M_2)} \times \frac{\pi(M_1)}{\pi(M_2)}, \tag{1.6}$$

where $m(\boldsymbol{y}^n; M_k)$ and $\pi(M_k)$ denote the marginal likelihood and prior probability for model M_k, respectively. The first term on the right side of (1.6) is known as the *Bayes factor* (BF) in favor of model M_1 when it is compared to M_2 (Kass and Raftery, 1995), i.e.,

$$\mathrm{BF}_{12} = \frac{m(\boldsymbol{y}^n; M_1)}{m(\boldsymbol{y}^n; M_2)}. \tag{1.7}$$

The BF can be interpreted as the summary of support provided by the data to one model as opposed to an alternative one. It can also be used for Bayesian hypothesis testing as originally suggested by Jeffreys (1935).

Since the marginal likelihood cannot be obtained analytically for many models, an alternative quantity based on the predictive density is

$$p(y_j|\boldsymbol{y}_{(-j)}, M_k) = \int p(y_j|\Theta_k, M_k)\, p(\Theta_k|\boldsymbol{y}_{(-j)}, M_k)d\Theta_k, \tag{1.8}$$

where, for $j = 1, \ldots, n$, $\boldsymbol{y}_{(-j)} = \{y_\ell | \ell = 1, \ldots, n; \ell \neq j\}$ and Θ_k denotes the parameters associated with model M_k. The above density evaluated at data point y_j is referred to as the *conditional predictive ordinate* (CPO). The pseudo likelihood obtained as the product of individual CPO's, i.e.,

$$\widehat{m}(\boldsymbol{y}^n) = \prod_{j=1}^n p(y_j | \boldsymbol{y}_{(-j)}, M_k) \qquad (1.9)$$

is called the *cross validation marginal likelihood*. Evaluation of the cross validation marginal likelihood typically requires Markov chain Monte Carlo (MCMC) simulation. Gelfand and Dey (1994) discuss how $p(y_j | \boldsymbol{y}_{(-j)}, M_k)$ can be approximated using posterior samples from the MCMC output. Once the cross validation marginal likelihood is approximated, the *pseudo Bayes factor* (PsBF) can be obtained to compare any two models.

Information criteria

As noted by Kass and Raftery (1995), the BF can be approximated using the *Schwarz criterion*, which is also known as the *Bayesian information criterion* (BIC). In the above, we can write the logarithm of the Bayes factor as

$$\log \mathrm{BF}_{12} = \log m(\boldsymbol{y}^n; M_1) - \log m(\boldsymbol{y}^n; M_2).$$

For large sample size n, it is possible to show that

$$\log m(\boldsymbol{y}^n; M_k) \approx \log L_k(\boldsymbol{y}^n; \widehat{\boldsymbol{\Theta}}_k) - (r_k/2) \log(n),$$

where $L_k(\boldsymbol{y}^n; \widehat{\boldsymbol{\Theta}}_k)$ is the likelihood for model M_k evaluated at the maximum likelihood estimator $\widehat{\boldsymbol{\Theta}}_k$ (or the posterior mode) of $\boldsymbol{\Theta}_k$, while r_k is the number of parameters for model M_k. In the above, $-2 \log m(\boldsymbol{y}^n; M_k)$ is known as the BIC for model M_k, i.e.,

$$\mathrm{BIC}_k = -2 \log L_k(\boldsymbol{y}^n; \widehat{\boldsymbol{\Theta}}_k) + r_k \log(n).$$

A smaller value of BIC_k implies more support for the model M_k. We can approximate BF using BIC as

$$\mathrm{BF}_{12} \approx \exp\Big[-\frac{1}{2}(\mathrm{BIC}_1 - \mathrm{BIC}_2)\Big].$$

Especially for hierarchical Bayesian models, it is not easy to evaluate BIC since the "effective" number of parameters in the model will not be known. For this purpose, the *deviance information criterion* (DIC) was proposed by Spiegelhalter et al. (2002). For a generic parameter vector $\boldsymbol{\Theta}$, the DIC is defined as

$$\mathrm{DIC} = \overline{D} + r_D, \text{ where,}$$
$$D = -2 \log L(\boldsymbol{\Theta}; \boldsymbol{y}^n), \quad \overline{D} = E_{\Theta | \boldsymbol{y}^n}(D), \text{ and}$$
$$r_D = \overline{D} - D(\widehat{\boldsymbol{\Theta}}), \qquad (1.10)$$

where $\widehat{\boldsymbol{\Theta}}$ is the posterior mean, $\overline{D}$ represents the "goodness of the fit" of the model, while the term r_D represents a complexity penalty as reflected by the *effective number of parameters* of the model. The unknown effective number of parameters of the model is estimated as the difference between the posterior mean of the deviance and the deviance evaluated at the posterior mean of Θ.

An alternative to *DIC* is the *Watanabe Akaike information criterion* (WAIC) which is defined as (see Watanabe and Opper (2010) and Gelman et al. (2014))

$$\text{WAIC} = \text{lppd} + p_{\text{WAIC}}, \tag{1.11}$$

where lppd is the log pointwise posterior density given by

$$\text{lppd} = \sum_{j=1}^{n} \log \left(\int p(y_j|\Theta_k)\pi(\Theta_k|\boldsymbol{y}^n)d\Theta_k \right).$$

The lppd is evaluated using draws $\Theta_k^s, s = 1, \ldots, S$ from the posterior distribution of Θ_k as

$$\sum_{j=1}^{n} \log \left(\frac{1}{S} \sum_{s=1}^{S} p(y_j|\Theta_k^s) \right),$$

and p_{WAIC} is a correction term for the effective number of parameters to adjust for overfitting,

$$p_{\text{WAIC}} = 2\sum_{j=1}^{n} \left(\log(E_{post}p(y_j|\Theta_k)) - E_{post}(\log p(y_j|\Theta_k)) \right),$$

which can be computed from draws $\Theta_k^s, s = 1, \ldots, S$ as

$$2\sum_{j=1}^{n} \left(\log \left(\frac{1}{S} \sum_{s=1}^{S} p(y_j|\Theta_k^s) \right) - \frac{1}{S}\sum_{s=1}^{S} \log p(y_j|\Theta_k^s) \right).$$

Model averaging

An alternative approach for predicting the future values of the response variable Y is to take into account model uncertainty and use Bayesian model averaging.

Assume that we have K possible models $M_k, \; k = 1, \ldots, K$ with prior model probabilities $p(M_k)$ such that $\sum_{k=1}^{K} p(M_k) = 1$. Typically, under model M_k, we have unknown parameter(s) Θ_k, and we denote the model for Y by $p(y|\Theta_k, M_k)$. Priors for Θ_k given M_k are denoted as $p(\Theta_k|M_k)$, for $k = 1, \ldots, K$.

If $\boldsymbol{y}^n$ denotes the observed data, then $p(y^*|\boldsymbol{y}^n)$, the posterior predictive distribution for a future value of Y, is given by

$$p(y^*|\boldsymbol{y}^n) = \sum_{k=1}^{K} p(y^*|\boldsymbol{y}^n, M_k)p(M_k|\boldsymbol{y}^n), \tag{1.12}$$

which is a weighted average of the posterior predictive distributions under each of the models. This is known as Bayesian model averaging (BMA). In (1.12), the weights are given by the posterior model probabilities

$$p(M_k|\boldsymbol{y}^n) = \frac{p(\boldsymbol{y}^n|M_k)\,p(M_k)}{\sum_{j=1}^{K} p(\boldsymbol{y}^n|M_j)\,p(M_j)}, \tag{1.13}$$

where the marginal likelihood term for model M_k is given by

$$p(\boldsymbol{y}^n|M_k) = \int p(\boldsymbol{y}^n|\boldsymbol{\theta}_k, M_k)\, p(\boldsymbol{\Theta}_k|M_k)d\boldsymbol{\Theta}_k. \qquad (1.14)$$

Finally, $p(y^*|\boldsymbol{y}^n, M_k)$ is obtained via

$$p(y^*|\boldsymbol{y}^n, M_k) = \int p(y^*|\boldsymbol{\Theta}_k, M_k)\, p(\boldsymbol{\Theta}_k|\boldsymbol{y}^n, M_k)d\boldsymbol{\Theta}_k, \qquad (1.15)$$

where

$$p(\boldsymbol{\Theta}_k|\boldsymbol{y}^n, M_k) \propto p(\boldsymbol{y}^n|\boldsymbol{\Theta}_k, M_k)p(\boldsymbol{\Theta}_k|M_k). \qquad (1.16)$$

1.3   Bayesian analysis of time series

In many applications, we are interested in modeling a time series of observations y_t, $t = 1, \ldots, n$, based on available information. The information might consist only of observed values of y_t, or it may include a set of past and current observations on exogenous predictors in addition to the past history of y_t. The observations $\boldsymbol{y}^n = (y_1, y_2, \ldots, y_n)'$, collected at discrete time points $t = 1, \ldots, n$, is a realization of the discrete time stochastic process $\{Y_t\}$.

The most important feature of the time series is possible serial correlation among the observations. Describing such correlation is the essence of time series modeling. As noted by Cox et al. (1981), such correlation is typically described by using either the *observation driven* or the *parameter driven* modeling strategies. The former includes models such as Autoregressive Moving Average (ARMA) processes and Markov chains, whereas the latter involves latent structure models such as the state-space models where the correlation is created by the time dependence among latent factors. The parameter driven models, due to their dynamic latent structure, are capable of capturing *nonstationary* behavior in the means and variances.

In our development, we will be dealing with Bayesian analysis of both the observation driven and parameter driven processes. The Bayesian approach to time series modeling is still driven by the notion of "describing uncertainty via probability" and adherence to the rules of probability and the likelihood principle. Typical Bayesian analysis of a time series model will still involve deriving posterior and predictive distributions of unknown quantities in the specified model. Due to the temporal dependence in time series data and the potential dynamic structure of model parameters, our inferences and predictions need to be performed sequentially. Sequential processing can be handled adequately in the Bayesian paradigm as will be discussed in what follows.

To be able to consider both the observation and parameter driven models, we will decompose our generic parameter vector Θ into static and dynamic parameters denoted by $\boldsymbol{\theta}$ and $\boldsymbol{x}^n = (x_1, \ldots, x_n)'$, respectively. Thus, we define $\Theta = (\boldsymbol{\theta}, \boldsymbol{x}^n)$. At time t, we denote the observed data by $\boldsymbol{y}^t = (y_1, y_2, \ldots, y_t)'$ and the posterior distribution of Θ by $\pi(\Theta|\boldsymbol{y}^t)$. It can be shown via Bayes' theorem, that

$$\pi(\Theta|\boldsymbol{y}^t) \propto \pi(\Theta|\boldsymbol{y}^{t-1})L(\Theta; y_t, \boldsymbol{y}^{t-1}), \qquad (1.17)$$

where $\pi(\Theta|\boldsymbol{y}^{t-1})$ is the posterior distribution of Θ at time $(t-1)$ which can be considered as the prior of Θ before observing y_t. The likelihood term $L(\Theta; y_t, \boldsymbol{y}^{t-1})$ is obtained by evaluating $p(y_t|\Theta, \boldsymbol{y}^{t-1})$ at the observed value y_t at time t. It is important to note that, for observation driven models, $p(y_t|\Theta, \boldsymbol{y}^{t-1}) \neq p(y_t|\Theta)$. Given $\boldsymbol{y}^t$, the predictive distribution of Y_{t+1} is given by

$$p(y_{t+1}|\boldsymbol{y}^t) = \int p(y^{t+1}|\Theta, \boldsymbol{y}^t)\pi(\Theta|\boldsymbol{y}^t)d\Theta. \qquad (1.18)$$

The marginal posterior distribution of the entire latent vector $\boldsymbol{x}^n$ given $\boldsymbol{y}^t$ is obtained by marginalization of the posterior distribution of Θ as

$$\pi(\boldsymbol{x}^n|\boldsymbol{y}^t) = \int \pi(\boldsymbol{x}^n, \boldsymbol{\theta}|\boldsymbol{y}^t)d\boldsymbol{\theta}.$$

The posterior distributions of the components of $\boldsymbol{x}^n$, i.e., $\pi(x_\tau|\boldsymbol{y}^t), \tau = 1, \dots, n$ can be obtained similarly. The distribution of the current latent component x_t, i.e., $\pi(x_t|\boldsymbol{y}^t)$ is referred to as the *filtering distribution*. The remaining marginals for x_τ, where $\tau < t$ and $\tau > t$ are known as the *smoothing distributions* and *forecast distributions*, respectively.

In sequential analysis of time series data, often the interest is centered around the filtering distribution of x_t which can be obtained from the sequential updating of x_t and $\boldsymbol{\theta}$ as

$$\pi(x_t, \boldsymbol{\theta}|\boldsymbol{y}^t) \propto \pi(x_t, \boldsymbol{\theta}|\boldsymbol{y}^{t-1})L(x_t, \boldsymbol{\theta}; y_t, \boldsymbol{y}^{t-1}).$$

The marginal posterior distribution of x_t obtained from $\pi(x_t, \boldsymbol{\theta}|\boldsymbol{y}^{t-1})$, i.e., $\pi(x_t|\boldsymbol{y}^{t-1})$, is the forecast distribution of x_t.

1.4 Gaussian dynamic linear models (DLMs)

In this section, we introduce Gaussian dynamic linear models (DLMs) for a univariate time series. Due to their ability to capture time varying parameters, the DLMs provide a flexible modeling framework for nonstationary time series. The Gaussian DLM leads to closed form expressions for estimates and forecasts, and may be the best path if such a model is suitable for the time series. We show a few examples that are often used in practice. In Chapter 3, we will fit these models to data using R-INLA (Martins et al., 2013).

One way to model a univariate time series y_t is as the sum of an unobservable (latent) level plus a random error. The latent level can either be constant over time or can randomly evolve over time according to some stochastic process. We start with a simple example for a univariate time series y_t, the *constant level plus noise model*, followed by a *local level model*, also referred to as the *random walk plus noise model*. In each case, we show by simple formulas how Bayesian updating is done. Next, we define the DLM framework for a univariate time series y_t and describe its properties, and present another commonly used example of DLM, the *AR(1) plus noise model*. We then give a brief review of a DLM for a vector-valued time series, and discuss the Kalman filtering and smoothing algorithms. We end the chapter with a brief look at how the DLM can be extended for non-Gaussian setups.

1.4.1 Constant level plus noise model

Consider a simple model for a univariate Gaussian time series y_t, which can be described as the sum of a constant level and a Gaussian random error (Shumway and Stoffer, 2017):

$$y_t = \alpha + v_t; \quad v_t \sim N(0, \sigma_v^2). \qquad (1.19)$$

We use this model to illustrate Bayesian updating for a time series. Suppose we assume that the prior distribution of the random variable α is normal with mean m_0 and variance C_0, i.e., $\pi(\alpha) \sim N(m_0, C_0)$, which is independent of the distribution of v_t for all t. For simplicity, we first assume that the error variance σ_v^2 is known. Given data $\boldsymbol{y}^n$, our objective is to update our opinion about α via the posterior distribution of α given $\boldsymbol{y}^n$:

$$\pi(\alpha|\boldsymbol{y}^n) \propto L(\alpha; \sigma_v^2, \boldsymbol{y}^n)\, \pi(\alpha),$$

where $L(\alpha; \sigma_v^2, \boldsymbol{y}^n)$ is a Gaussian likelihood.

Due to the conjugacy of the normal likelihood and the normal prior for α, the posterior distribution of α given $\boldsymbol{y}^n$ is normal, i.e.,

$$\pi(\boldsymbol{\alpha}|\boldsymbol{y}^n) \sim N(m_n, C_n), \text{ where,}$$

$$m_n = E(\alpha|\boldsymbol{y}^n) = \frac{C_0}{C_0 + \frac{\sigma_v^2}{n}} \overline{y} + \frac{\frac{\sigma_v^2}{n}}{C_0 + \frac{\sigma_v^2}{n}} m_0, \text{ and}$$

$$C_n = Var(\alpha|\boldsymbol{y}^n) = \left(\frac{n}{\sigma_v^2} + \frac{1}{C_0} \right)^{-1} = \frac{\sigma_v^2 C_0}{\sigma_v^2 + nC_0}.$$

Note that m_n is a weighted average of the sample mean $\overline{y} = \frac{1}{n}\sum_{t=1}^{n} y_t$ and the prior mean m_0, where the weight is a function of C_0 and σ_v^2. If C_0 is large, then the posterior mean m_n is approximately equal to $\overline{Y}$ and $C_n \approx \sigma_v^2/n$. The posterior precision $1/C_n$ is given by

$$\frac{1}{C_n} = \frac{n}{\sigma_v^2} + \frac{1}{C_0},$$

where n/σ_v^2 denotes the precision of the sample mean while $1/C_0$ is the prior precision. Note that the posterior precision $1/C_n$ is always larger than the prior precision $1/C_0$.

It is easy to see how the posterior distribution of α can be obtained sequentially over time. At time $t = n - 1$,

$$\pi(\alpha|\boldsymbol{y}^{n-1}) \sim N(m_{n-1}, C_{n-1}).$$

This distribution then plays the role of the prior for α which is updated, when y_n is observed at time $t = n$, to yield the posterior distribution

$$\pi(\alpha|\boldsymbol{y}^n) \sim N(m_n, C_n), \text{ where,}$$

$$m_n = \frac{C_{n-1}}{C_{n-1} + \sigma_v^2} y_n + \left(1 - \frac{C_{n-1}}{C_{n-1} + \sigma_v^2} \right) m_{n-1}, \text{ and}$$

$$C_n = \left(\frac{1}{\sigma_v^2} + \frac{1}{C_{n-1}} \right)^{-1} = \frac{\sigma_v^2 C_{n-1}}{\sigma_v^2 + C_{n-1}}.$$

We can obtain the predictive distribution of y_{n+1} given the information up to time n as

$$y_{n+1}|\boldsymbol{y}^n \sim N(m_n, C_n + \sigma_v^2).$$

Note that m_n is the posterior mean of α and the one-step ahead point prediction of y_{n+1}. This leads to an *error correction form* representation of the prediction as

$$m_n = m_{n-1} + \frac{C_{n-1}}{C_{n-1} + \sigma_v^2}(y_n - m_{n-1}).$$

Thus, m_n is obtained by correcting the previous estimate m_{n-1} with a weighted forecast error $e_n = y_n - m_{n-1}$, the weight being

$$\frac{C_{n-1}}{C_{n-1} + \sigma_v^2} = \frac{C_0}{\sigma_v^2 + nC_0}.$$

This model is too simplistic to offer a realistic fit to time series in practice. Therefore, we do not discuss this further in the book.

1.4.2 Local level model

In many situations, we may wish to model the level as dynamically evolving over time. One assumption is that the level evolves as a random walk. For this reason, this model is also called a random walk plus noise model:

$$y_t = x_t + v_t; \quad v_t \sim N(0, \sigma_v^2), \tag{1.20}$$

$$x_t = x_{t-1} + w_t; \quad w_t \sim N(0, \sigma_w^2). \tag{1.21}$$

Here, v_t and w_t are random errors which are assumed to follow zero mean normal distributions with unknown variances σ_v^2 and σ_w^2 respectively. The model in (1.20) and (1.21) is one of the simplest *dynamic* models for a univariate time series y_t and is appropriate when the time series exhibits a slowly changing level over time. We illustrate how the dynamic component impacts Bayesian inference of the latent variable x_t, before and after we observe the data y_t at time t.

Prior distribution of x_t. After we see y_{t-1} at time $t-1$, we can obtain the posterior distribution of x_{t-1} given $\boldsymbol{y}^{t-1} = (y_1, \ldots, y_{t-1})$, i.e., $\pi(x_{t-1}|\boldsymbol{y}^{t-1})$. This leads to the prior distribution for x_t before we observe y_t; i.e., $\pi(x_t|\boldsymbol{y}^{t-1})$.

Forecast distribution of y_t. At time $t-1$, we predict the output y_t for time t using $p(y_t|\boldsymbol{y}^{t-1})$.

Posterior distribution of x_t. Once the data at time t, i.e., y_t, becomes available, we update our belief about x_t, i.e., we obtain the posterior of x_t given $\boldsymbol{y}^t$, denoted by $\pi(x_t|\boldsymbol{y}^t)$. As previously mentioned, this is the filtering distribution for x_t.

We next illustrate these three steps. At time $t-1$, suppose that

$$\pi(x_{t-1}|\boldsymbol{y}^{t-1}) \sim N(m_{t-1}, C_{t-1}).$$

Using the state equation

$$x_t = x_{t-1} + w_t; \quad w_t \sim N(0, \sigma_w^2),$$

we can obtain the forecast distribution of x_t as

$$\pi(x_t|\boldsymbol{y}^{t-1}) \sim N(a_t, R_t),$$

where

$$a_t = m_{t-1}, \text{ and } R_t = C_{t-1} + \sigma_w^2.$$

Since $\sigma_w^2 > 0$, we become *more uncertain* about the average level x_t than under the constant level model. Standing at time $t - 1$, we can also predict the next observation y_t using the predictive distribution

$$y_t | \boldsymbol{y}^{t-1} \sim N(f_t, Q_t), \text{ where}$$

$$f_t = a_t \text{ and } Q_t = R_t + \sigma_v^2.$$

The uncertainty about y_t depends on the measurement error σ_v^2 as well as on the uncertainty about the average level x_t as reflected by the state equation.

At time t, the new observation y_t becomes available, and we update our uncertainty about x_t with the new measurement y_t. We obtain the posterior density $\pi(x_t | \boldsymbol{y}^t)$, i.e., the *filtering density* as

$$x_t | \boldsymbol{y}^t \sim N(m_t, C_t),$$

where

$$m_t = a_t + \frac{R_t}{R_t + \sigma_v^2}(y_t - f_t)$$

and

$$C_t = \frac{\sigma_v^2 R_t}{\sigma_v^2 + R_t}.$$

Note that for this step, (i) the prior distribution of x_t is $N(a_t, R_t)$ and (ii) y_t is independent of its past history given x_t. This sequential updating mechanism is known as *estimation-correction structure*, because the previous best estimate a_t is corrected by a fraction of the forecast error $e_t = y_t - f_t$ having weight $K_t = R_t/(R_t + \sigma_v^2)$. The magnitude of the state error variance σ_w^2 relative to the observation error variance σ_v^2 is known as the *signal-to-noise ratio (SNR)*. This quantity plays a crucial role in determining the effect of data on estimation and forecasting.

1.4.3   Gaussian DLM framework for univariate time series

The Gaussian DLM is a general and flexible framework for analyzing many different classes of time series (Shumway and Stoffer, 2017). The DLM consists of an observation equation and a state equation, which are respectively given by

$$y_t = \alpha + F_t x_t + v_t, \ v_t \sim N(0, \sigma_v^2), \qquad (1.22)$$

and

$$x_t = G_t x_{t-1} + w_t, \ w_t \sim N(0, \sigma_w^2). \qquad (1.23)$$

The observation equation expresses y_t as a linear combination of the latent variable x_t and an observation error v_t plus a constant level α; here F_t is known. The state equation expresses x_t as a linear combination of x_{t-1} and a state error w_t, where G_t represents the transition from x_{t-1} to x_t and may be known or unknown or time-varying or constant. We assume that the initial state is Gaussian, i.e., $x_0 \sim N(m_0, C_0)$.

Usually, α, σ_v^2, σ_w^2, m_0, and C_0 will be unknown and must be estimated from the data. Let $\boldsymbol{\theta}$ denote the collection of all the unknown scalars that we refer to as hyperparameters, i.e., $\boldsymbol{\theta} = (\alpha, \sigma_v^2, \sigma_w^2, m_0, C_0)'$. In the DLM, the primary *parameter* of interest is the latent process x_t.

Remark 1. The latent process x_t is a *Markov chain*. This means that the conditional distribution of x_t given its entire history, $x_{t-1}, x_{t-2}, x_{t-3}, \ldots$ only depends on the previous state x_{t-1}, i.e.,

$$\pi(x_t | x_{t-1}, x_{t-2}, x_{t-3}, \ldots) = \pi(x_t | x_{t-1}).$$

The distribution may depend on the hyperparameters $\boldsymbol{\theta}$.

Remark 2. Conditionally on x_t, the observed data y_t is independent of its own history $y_{t-1}, y_{t-2}, y_{t-3}, \ldots$. All the information about the history of y_t is included in the current state x_t, which then determines y_t.

Remark 3. At each time t, we can explain y_t as a linear function of x_t and an additive Gaussian noise. The latent process x_t follows the Markovian dynamics in a linear way; i.e., x_t depends *linearly* only on x_{t-1} and an additive Gaussian noise w_t.

For example, the random walk plus noise (or local level) model defined in (1.20) and (1.21) for a univariate time series y_t is a special case of (1.22) and (1.23) with $F_t = 1$, $G_t = 1$ and $\alpha = 0$. Further, if we set $\sigma_w^2 = 0$, and $\alpha \neq 0$, the random walk plus noise model reduces to the constant plus noise model where α denotes a constant level that does not vary over time. In the next subsection, we show another simple and widely used example of a DLM.

1.4.4 AR(1) plus noise model

Rather than a random walk assumption for the dynamic evolution of the state variable in (1.23), we can allow the level x_t to evolve over time as an autoregression of order 1, i.e., an AR(1) process with serial correlation ϕ. That is, in (1.23), we set $G_t = \phi$ instead of $G_t = 1$ for all t. The AR(1) plus noise model is

$$y_t = x_t + v_t; \quad v_t \sim N(0, \sigma_v^2), \tag{1.24}$$

$$x_t = \phi x_{t-1} + w_t; \quad w_t \sim N(0, \sigma_w^2). \tag{1.25}$$

For process stability, it is usual to assume that $|\phi| < 1$. The other assumptions are similar to those under the random walk plus noise model. We discuss analyzing and forecasting univariate time series under these DLMs in Chapter 3. We end this section by introducing the DLM for a vector-valued process.

1.4.5 DLM for vector-valued time series

Let $\boldsymbol{y}_t$ be a q-dimensional observed time series, for $q \geq 1$ and let $\boldsymbol{x}_t$ be a p-dimensional unobserved (or latent) state process for $p \geq 1$. The DLM for $\boldsymbol{y}_t$ is described through the observation and state equations shown below:

$$\boldsymbol{y}_t = \boldsymbol{F}_t \boldsymbol{x}_t + \boldsymbol{\Gamma} \boldsymbol{u}_{1,t} + \boldsymbol{v}_t, \quad \boldsymbol{v}_t \sim N_q(\boldsymbol{0}, \boldsymbol{V}), \tag{1.26}$$

and

$$\boldsymbol{x}_t = \boldsymbol{G}_t \boldsymbol{x}_{t-1} + \boldsymbol{\Upsilon} \boldsymbol{u}_{2,t} + \boldsymbol{w}_t, \ \boldsymbol{w}_t \sim N_p(\boldsymbol{0}, \boldsymbol{W}). \tag{1.27}$$

In (1.26) and (1.27), for $t = 1, \ldots, n$,

- $\boldsymbol{F}_t$ is a known matrix of order $q \times p$,

- $\boldsymbol{u}_{j,t}$ are known r_j-dimensional vectors of exogenous predictors, for $j = 1, 2$,

- $\boldsymbol{\Gamma}$ is a $q \times r_1$ matrix of regression coefficients corresponding to $\boldsymbol{u}_{1,t}$,

- $\boldsymbol{\Upsilon}$ is a $p \times r_2$ matrix of regression coefficients corresponding to $\boldsymbol{u}_{2,t}$,

- $\boldsymbol{G}_t$ is a possibly unknown state transition matrix of order $p \times p$,

- $\boldsymbol{v}_t \sim N_q(\boldsymbol{0}, \boldsymbol{V})$ is the observation error process,

- $\boldsymbol{w}_t \sim N_p(\boldsymbol{0}, \boldsymbol{W})$ is the state error process,

- $\{\boldsymbol{v}_t\}$ and $\{\boldsymbol{w}_t\}$ are mutually independent white noise processes,

- the initial state $\boldsymbol{x}_0 \sim N_p(\boldsymbol{m}_0, \boldsymbol{C}_0)$, and

- $\boldsymbol{x}_0$ is independent of $\{\boldsymbol{v}_t\}$ and $\{\boldsymbol{w}_t\}$.

We write the DLM properties in terms of a general distributional setup. As we mentioned in the univariate case, here too, the state process is assumed to be linear, Markovian, and Gaussian, i.e.,

$$\pi(\boldsymbol{x}_t | \boldsymbol{x}_{t-1}, \boldsymbol{x}_{t-2}, \ldots) = \pi(\boldsymbol{x}_t | \boldsymbol{x}_{t-1}) = f_w(\boldsymbol{x}_t - \boldsymbol{G}_t \boldsymbol{x}_{t-1} - \boldsymbol{\Upsilon} \boldsymbol{u}_{2,t}), \tag{1.28}$$

where $f_w(.)$ denotes a p-variate Gaussian p.d.f. with mean vector $\boldsymbol{0}$ and covariance matrix $\boldsymbol{W}$. The observations are linear, Gaussian, and conditionally independent given the state vector, i.e.,

$$p(\boldsymbol{y}_t | \boldsymbol{x}_t, \boldsymbol{y}_{t-1}, \ldots, \boldsymbol{y}_1) = p(\boldsymbol{y}_t | \boldsymbol{x}_t) = f_v(\boldsymbol{y}_t - \boldsymbol{F}_t \boldsymbol{x}_t - \boldsymbol{\Gamma} \boldsymbol{u}_{1,t}), \tag{1.29}$$

where $f_v(.)$ is a q-variate Gaussian p.d.f. with mean vector $\boldsymbol{0}$ and covariance matrix $\boldsymbol{V}$. The joint distribution of the observation and state processes is specified by

$$\pi(\boldsymbol{x}_0, \boldsymbol{x}_1, \ldots, \boldsymbol{x}_n, \boldsymbol{y}_1, \ldots, \boldsymbol{y}_n) = \pi(\boldsymbol{x}_0) \times \prod_{t=1}^{n} f_w(\boldsymbol{x}_t - \boldsymbol{G}_t \boldsymbol{x}_{t-1} - \boldsymbol{\Upsilon} \boldsymbol{u}_{2,t}) f_v(\boldsymbol{y}_t - \boldsymbol{F}_t \boldsymbol{x}_t - \boldsymbol{\Gamma} \boldsymbol{u}_{1,t}).$$

$$\tag{1.30}$$

1.4.6   Kalman filtering and smoothing

Kalman filtering and smoothing for the Gaussian DLM have been described and implemented in R packages such as `dlm` (Petris, 2010) and `astsa` (Shumway and Stoffer, 2017). At each time t, we can estimate the latent state vector $\boldsymbol{x}_t$ given data $\boldsymbol{Y}^s = (\boldsymbol{y}_1', \boldsymbol{y}_2', \ldots, \boldsymbol{y}_s')'$, as $\boldsymbol{x}_t^s$, where $s < t$ or $s = t$ or $s > t$. Differences in the information set $\boldsymbol{Y}^s$ on which we condition lead to the three different situations listed below:

- $s < t$: Forecasting/prediction. For example, when $s = t - 1$, we estimate $\boldsymbol{x}_t$ by $\boldsymbol{x}_t^{t-1} = E(\boldsymbol{x}_t | \boldsymbol{Y}^{t-1})$, its one-step ahead forecast estimate;

- $s = t$: Filtering. When $s = t$, we obtain the filter estimate of $\boldsymbol{x}_t$ as $\boldsymbol{x}_t^t = E(\boldsymbol{x}_t | \boldsymbol{Y}^t)$;

- $s > t$: Smoothing. For example, when $s = n > t$, we estimate $\boldsymbol{x}_t$ by $\boldsymbol{x}_t^n = E(\boldsymbol{x}_t | \boldsymbol{Y}^n)$, its smoothed estimate.

Note that the `R-INLA` implementation of a DLM produces the smoothed estimate of the states, i.e., $\boldsymbol{x}_t^n$ and not the filter estimates $\boldsymbol{x}_t^t$ for $t = 1, \ldots, n$. In fact, as we show in Chapter 3, one must iteratively run the `inla()` function n times in order to obtain the filter estimates of $\boldsymbol{x}_t$.

1.5 Beyond basic Gaussian DLMs

So far in this chapter, we have seen a few examples of Gaussian DLMs. In Chapters 3 and 4, we will describe how to use `R-INLA` to fit these models to time series exhibiting various characteristics. The statistical details of the INLA framework itself are briefly reviewed in Chapter 2. In Chapter 5, we show how to fit Gaussian DLMs to time series when we have exogenous predictors observed over the same time frame as the time series we wish to model and predict, also including structural modeling (Harvey and Koopman, 2014).

Hierarchical dynamic linear models (HDLMs) allow us to model a set of g independent time series, each of length n (Gamerman and Migon, 1993). HDLMs are often referred to as panel time series models in the literature. In Chapter 6, we consider models for a set of g univariate time series.

While Gaussian DLMs are relatively easy to fit to data, they may not always be suitable for various types of complex time series data from different application domains. Examples consist of positive, continuous-valued time series, binary-valued time series, or count time series (West and Harrison, 1997). Chapter 7 describes models for time series following gamma, Weibull or beta distributions, while in Chapter 8, we describe dynamic model fitting for binary-valued time series as well as categorical time series. In Chapter 9, we use `R-INLA` to model and forecast univariate count time series, as well as sets of univariate count time series.

A popular example of a nonlinear and non-Gaussian dynamic time series model is the stochastic volatility (SV) model that is used for modeling financial log returns, for which a Bayesian framework was described in Kim et al. (1998). The SV model is also useful in other application domains to model logarithms of growths (which are similar to log returns). In Chapter 10, we describe dynamic stochastic volatility models using `R-INLA`. Chapter 11 discusses spatio-temporal modeling.

Chapter 12 describes DLMs for q-dimensional time series and approaches for forecasting them. A mitigating factor here is that `R-INLA` is currently only able to handle dimensions $q \leq 5$. Chapter 13 extends the hierarchical models in Chapter 6 to multivariate time series, using ideas from Chapter 12. The chapter also considers level correlated models for hierarchical modeling of multivariate count time series. The use of INLA in these situations offers a useful computationally feasible alternative to MCMC methods which can be too slow to be practically useful in many situations.

Chapter 1 – Appendix

Conditional distributions

We review some useful notation and formulas. Let W, Y, and Z denote random variables.

1. Conditional distribution of Y given Z:

$$p(Y|Z) = \frac{p(Y,Z)}{p(Z)}, \tag{1.31}$$

provided the denominator $p(Z)$ is positive. We can write (1.31) equivalently as

$$p(Z) = \frac{p(Y,Z)}{p(Y|Z)}. \tag{1.32}$$

2. Conditioning both sides of (1.32) on W,

$$p(Z|W) = \frac{p(Y,Z|W)}{p(Y|Z,W)}. \tag{1.33}$$

We can rewrite (1.33) equivalently as

$$p(Y,Z|W) = p(Z|W) \times p(Y|Z,W). \tag{1.34}$$

Exponential family of distributions

One-parameter exponential family

The one-parameter exponential family is defined by the p.m.f. or p.d.f.

$$p(y|\theta) = h(y) \exp\left[\eta(\theta)t(y) - b(\theta)\right], \tag{1.35}$$

where $t(y), h(y), b(\theta)$ are known functions. The support of Y does not depend on θ. The distribution (1.35) can also be written as

$$p(y|\theta) = h(y)\, g(\theta) \exp\left[\eta(\theta)\, t(y)\right], \tag{1.36}$$

or

$$p(y|\theta) = \exp\left[\eta(\theta)\, t(y) - b(\theta) + c(y)\right]. \tag{1.37}$$

If $\eta(\theta) = \theta$, the exponential family is said to be in canonical form.

k-parameter exponential family

Let $\boldsymbol{\theta}$ be a k-dimensional vector. The exponential family p.m.f. or p.d.f. has the form

$$p(y|\boldsymbol{\theta}) = h(y) \exp\left[\sum_{i=1}^{k} \eta_i(\boldsymbol{\theta})\, t_i(y) - b(\boldsymbol{\theta})\right], \tag{1.38}$$

where $\eta_i(\boldsymbol{\theta})$, $i = 1, \ldots, k$ are real-valued functions of only $\boldsymbol{\theta}$ and $t_i(y)$, $i = 1, \ldots, k$ are real-valued functions of only y. If $\eta_i(\boldsymbol{\theta}) = \theta_i$, for all i, the k-parameter exponential family will be in canonical form. We can also write (1.38) as

$$p(y|\boldsymbol{\theta}) = h(y) \exp\left[\boldsymbol{\eta}(\boldsymbol{\theta})' \, \boldsymbol{t}(y) - b(\boldsymbol{\theta})\right], \tag{1.39}$$

or as

$$p(y|\boldsymbol{\theta}) = h(y) \, g(\boldsymbol{\theta}) \exp\left[\boldsymbol{\eta}(\boldsymbol{\theta})' \, \boldsymbol{t}(y)\right], \tag{1.40}$$

where $\boldsymbol{\eta}(\boldsymbol{\theta})' = \left(\eta_1(\boldsymbol{\theta}), \ldots, \eta_k(\boldsymbol{\theta})\right)$ and $\boldsymbol{t}(y) = \left(t_1(y), \ldots, t_k(y)\right)$.

Exponential dispersion family of distributions

Let θ be a scalar parameter of interest. Let $\phi > 0$ be a *dispersion parameter* which may or may not be known. Consider the p.d.f.

$$p(y|\theta, \phi) = \exp\left[\frac{y\theta - b(\theta)}{\phi} + c(y, \phi)\right], \tag{1.41}$$

where $b(\cdot)$ and $c(\cdot, \cdot)$ are specified functions. Here, θ is called the *canonical parameter* of the distribution, $b(\theta)$, ϕ and $c(y, \phi)$ are related by the condition that $p(y|\theta, \phi)$ must integrate to 1. The function $b(\theta)$ is usually explicitly given for standard distributions, while $c(y, \phi)$ is left implicit. This is not a problem for estimating θ, since the score equation does not involve $c(y, \phi)$. However, without an explicit $c(y, \phi)$, likelihood based estimation of the dispersion ϕ, or a likelihood based inference for (θ, ϕ) are not possible.

The log-likelihood function of (θ, ϕ) is given by

$$\ell(\theta, \phi; y) = \log p(y|\theta, \phi) = \frac{\left[y\theta - b(\theta)\right]}{\phi} + c(y, \phi). \tag{1.42}$$

To find the mean and variance of $Y \sim p(y|\theta, \phi)$, recall that

$$E[\nabla \ell(\theta, \phi; y)] = E(\partial\ell/\partial\theta) = 0, \tag{1.43}$$
$$E(\partial^2\ell/\partial\theta^2) + E(\partial\ell/\partial\theta)^2 = 0, \tag{1.44}$$

where

$$\partial\ell/\partial\theta = \{y - \partial b(\theta)/\partial\theta\}/\phi = \{y - b'(\theta)\}/\phi,$$
$$\partial^2\ell/\partial\theta^2 = -\{\partial^2 b(\theta)/\partial\theta^2\}/\phi = -b''(\theta)/\phi,$$

for the p.d.f. (1.41). It follows from the above that

$$E(Y) = \mu = \partial b(\theta)/\partial\theta = b'(\theta), \text{ and} \tag{1.45}$$
$$Var(Y) = E\left[(Y - \partial b(\theta)/\partial\theta)^2\right] = \partial^2 b(\theta)/\partial\theta^2 \phi = \phi b''(\theta). \tag{1.46}$$

Since $\phi > 0$, as long as Y is nondegenerate, $b''(\theta) > 0$, so that b is strictly convex and b' is strictly increasing. Then the correspondence (1.45) is 1-to-1, and the distribution of Y can be parameterized by $\mu = E(Y)$ and ϕ via

$$\theta = (b')^{-1}(\mu). \tag{1.47}$$

The function $b''(\theta)$ depends on the canonical parameter, and hence on μ, and is called the variance function, and is denoted by $v(\mu)$. Then,

$$Var(Y) = \phi v(\mu) \text{ with } v(\mu) = b''((b')^{-1}(\mu)). \qquad (1.48)$$

This explains why ϕ is called the dispersion parameter in a generalized linear model (GLIM). It plays a role similar to σ^2.

2

A Review of INLA

2.1 Introduction

This chapter gives a brief review of the Integrated Nested Laplace approximation (INLA), which was a framework proposed by Rue et al. (2009) for approximate Bayesian inference in a subclass of structured additive regression models (Fahrmeir and Tutz, 2001), called *latent Gaussian models*. Latent Gaussian models are a subset of Bayesian additive models with a structured additive predictor for modeling a response vector $\boldsymbol{y}^n = (y_1, y_2, \ldots, y_n)'$. Let $\boldsymbol{x}^n = (x_1, x_2, \ldots, x_n)'$ denote the vector of all the latent Gaussian variables, and $\boldsymbol{\theta}$ denote a vector of hyperparameters, which are not necessarily Gaussian. In latent Gaussian models, the response variable can belong to an exponential family of distributions, including the normal, binomial, Poisson, negative binomial, etc. (see Chapter 1 – Appendix), the mean of the response is linked to structured additive predictors, and hyperparameters enable us to handle variance terms in the latent Gaussian model or the dispersion parameter in the negative binomial response model, etc. INLA approaches the modeling via a hierarchical structure and can directly compute very accurate approximations of posterior densities which significantly decreases the computational time compared to fully Bayesian inference via MCMC.

Since the Gaussian Markov random field (GMRF) structure is used in INLA for handling structural time series models, we give a brief description in Chapter 2 – Appendix. While the level of detail given in this chapter may not be necessary for a user who is primarily interested in using R-INLA for data analysis, it does facilitate an understanding of how INLA works. For more details, see Lauritzen (1996) or Rue and Held (2005). Since the Laplace approximation plays an important role in the implementation of INLA, we next give an overview of Laplace's method as well as its modified version that is used in INLA.

2.2 Laplace approximation

The Laplace approach is used to approximate integrals which cannot be evaluated analytically. A general review of Laplace's method, which is based on asymptotic expansions, can be found in the text by De Bruijn (1981). The approach was originally considered in Bayesian statistics by Lindley (1961), but has gained more attention with Lindley (1980), who proposed its use for approximating posterior moments.

Consider the ratio of integrals given by

$$\frac{\int w(\Theta)\, e^{\ell(\Theta)} d\Theta}{\int \pi(\Theta)\, e^{\ell(\Theta)} d\Theta}, \tag{2.1}$$

where $\ell(\Theta) = \ell(\Theta; \boldsymbol{y}^n)$ is the log-likelihood function of the p-dimensional parameter vector Θ based on n observations $\boldsymbol{y}^n = (y_1, \ldots, y_n)'$ coming from the probability model $p(\boldsymbol{y}^n | \Theta)$, i.e.,

$$\ell(\Theta) = \sum_{t=1}^{n} \log p(y_t | \Theta).$$

The quantities $w(\Theta)$ and $\pi(\Theta)$ are functions of Θ which may need to satisfy certain conditions depending on the context.

In the Bayesian formulation, $\pi(\Theta)$ is the prior and $w(\Theta) = u(\Theta)\pi(\Theta)$, where $u(\Theta)$ is some function of Θ which is of interest. Thus, the ratio represents the posterior expectation of $u(\Theta)$, i.e., $E(u(\Theta)|\boldsymbol{y}^n)$. For example, if $u(\Theta) = \Theta$, then the ratio in (2.1) gives us the posterior mean of Θ. Similarly, if $u(\Theta) = p(\boldsymbol{y}^n | \Theta)$, it yields the posterior predictive distribution value at $\boldsymbol{y}^n$. Alternatively, we can write the ratio in (2.1) as

$$\frac{\int u(\Theta) e^{\Lambda(\Theta)} d\Theta}{\int e^{\Lambda(\Theta)} d\Theta}, \tag{2.2}$$

where
$$\Lambda(\Theta) = \ell(\Theta) + \log \pi(\Theta).$$

Lindley (1980) developed asymptotic expansions for the ratio of integrals in (2.2) as the sample size n gets large. The idea is to obtain a Taylor series expansion of all the above functions of Θ about $\widehat{\Theta}$, the posterior mode. Lindley's approximation to $E[u(\Theta)|\boldsymbol{y}^n]$ is given by

$$E[u(\Theta)|\boldsymbol{y}^n] \approx u(\widehat{\Theta}) + \frac{1}{2}\left(\sum_{i,j=1}^{p} u_{i,j} \ h_{i,j} + \sum_{i,j,k,l=1}^{p} \Lambda_{i,j,k} u_l \ h_{i,j} \ h_{k,l} \right), \tag{2.3}$$

where

$$u_i \equiv \frac{\partial u(\Theta)}{\partial \Theta_i} \bigg|_{\Theta^n = \widehat{\Theta}}, \ u_{i,j} \equiv \frac{\partial^2 u(\Theta)}{\partial \Theta_i \partial \Theta_j} \bigg|_{\Theta^n = \widehat{\Theta}}, \ \Lambda_{i,j,k} \equiv \frac{\partial^3 \Lambda(\Theta)}{\partial \Theta_i \partial \Theta_j \partial \Theta_k} \bigg|_{\Theta^n = \widehat{\Theta}},$$

and $h_{i,j}$ are the elements of the negative of the inverse Hessian of Λ at $\widehat{\Theta}$.

Lindley's approximation (2.3) involves third order differentiation and therefore is computationally cumbersome in highly parameterized cases. Tierney and Kadane (1986) proposed an alternative approximation which involves only the first and second order derivatives. This is achieved by using the mode of the product $u(\Theta)e^{\Lambda(\Theta)}$ rather than the mode of the posterior $e^{\Lambda(\Theta)}$ and evaluating the second derivatives at this mode. The Tierney-Kadane method approximates $E[u(\Theta)|\boldsymbol{y}^n]$ by

$$E[u(\Theta)|\boldsymbol{y}^n] \approx \left(\frac{|\Sigma^*(\widehat{\Theta})|}{|\Sigma(\widehat{\Theta})|} \right)^{1/2} \exp\left[n\left(\Lambda^*(\widehat{\Theta}) - \Lambda(\widehat{\Theta}) \right) \right], \tag{2.4}$$

where
$$\Lambda^*(\Theta) = \log u(\Theta) + \Lambda(\Theta),$$

and $\Sigma^*(\widehat{\Theta})$ and $\Sigma(\widehat{\Theta})$ are the corresponding negative inverse Hessians of Λ^* and Λ evaluated at $\widehat{\Theta}$, respectively.

2.2.1 Simplified Laplace approximation

Consider the Tierney-Kadane setup where we are interested in $\pi(\Theta_i|\boldsymbol{y}^n)$, the posterior marginal of Θ_i given $\boldsymbol{y}^n$. In other words, we are interested in the ratio of integrals

$$\frac{\int \pi(\Theta)e^{\ell(\Theta)}d\Theta_{(-i)}}{\int \pi(\Theta)e^{\ell(\Theta)}d\Theta},$$

where $\Theta_{(-i)} = (\Theta_j|j = 1,\ldots n; j \neq i)$. Tierney and Kadane (1986) provide the Laplace approximation for $\pi(\Theta_i|\boldsymbol{y}^n)$ as

$$\pi(\Theta_i|\boldsymbol{y}^n) \approx \left(\frac{|\boldsymbol{\Sigma}^*(\Theta_i)|}{2\pi n|\boldsymbol{\Sigma}(\widehat{\Theta}^n)|}\right)^{1/2} \frac{\pi(\Theta_i,\widehat{\Theta}_{(-i)})e^{\ell(\Theta_i,\widehat{\Theta}_{(-i)})}}{\pi(\widehat{\Theta}^n)e^{\ell(\widehat{\Theta}^n)}}, \tag{2.5}$$

where $\widehat{\Theta}_{(-i)} = \widehat{\Theta}_{(-i)}(\Theta_i)$ give the maxima of $\pi(\Theta_i,\Theta_{(-i)})e^{\ell(\Theta_i,\Theta_{(-i)})}$ for fixed Θ_i, and $\boldsymbol{\Sigma}^*(\Theta_i)$ is the corresponding negative inverse Hessian evaluated at $\widehat{\Theta}_{(-i)}$. As before, $\boldsymbol{\Sigma}(\widehat{\Theta})$ is the negative inverse Hessian of the log posterior of Θ evaluated at the posterior mode $\widehat{\Theta}$.

Denoting the joint distribution of Θ and $\boldsymbol{y}^n$ by $\pi(\Theta,\boldsymbol{y}^n)$, we can write

$$\pi(\Theta_i|\boldsymbol{y}^n) \propto \frac{\pi(\Theta,\boldsymbol{y}^n)}{\pi(\Theta_{(-i)}|\Theta_i,\boldsymbol{y}^n)} = \pi(\Theta_i,\boldsymbol{y}^n).$$

Rue et al. (2009) approximate the above as proportional to

$$\pi_{LA}(\Theta_i|\boldsymbol{y}^n) \propto \frac{\pi(\Theta,\boldsymbol{y}^n)}{\pi_G(\Theta_{(-i)}|\Theta_i,\boldsymbol{y}^n)}\bigg|_{\Theta_{(-i)}=\widehat{\Theta}_{(-i)}}$$

where $\pi_G(\Theta_{(-i)}|\Theta_i,\boldsymbol{y}^n)$ is called a *Gaussian approximation* to the density $\pi(\Theta_{(-i)}|\Theta_i,\boldsymbol{y}^n)$. More specifically,

$$\pi_G(\Theta_{(-i)}|\Theta_i,\boldsymbol{y}^n) = N(\widehat{\Theta}_{(-i)},\boldsymbol{\Sigma}^*(\Theta_i)).$$

As noted by the authors, $\pi_{LA}(\Theta_i|\boldsymbol{y}^n)$ is equivalent to Tierney and Kadane's posterior approximation (2.5). This is easy to see since the above will reduce to

$$\pi_{LA}(\Theta_i|\boldsymbol{y}^n) \propto \pi(\Theta_i,\widehat{\Theta}_{(-i)},\boldsymbol{y}^n)|\boldsymbol{\Sigma}^*(\Theta_i)|^{1/2}, \tag{2.6}$$

which is equivalent to (2.5).

Since the Laplace approximation (2.6) is computationally cumbersome when n is large due to the evaluation of a Hessian for each i, Rue et al. (2009) proposed a simplified version of the approximation. The proposed approach evaluates $\pi_G(\Theta_{(-i)}|\Theta_i,\boldsymbol{y}^n)$ in (2.6) using the conditional mean of $\Theta_{(-i)}$ given Θ_i implied by the Gaussian approximation to $\pi(\Theta|\boldsymbol{y}^n)$, the complete posterior of Θ. The resulting approximation is referred to as the *Simplified Laplace Approximation* (SLA) by the authors. Since the conditional variance of $\Theta_{(-i)}$ will not depend on Θ_i in the Gaussian case, the Hessian will be constant, i.e., the SLA is given by

$$\pi_{SLA}(\Theta_i|\boldsymbol{y}^n) \propto \pi(\Theta_i,\widehat{\Theta}_{(-i)},\boldsymbol{y}^n). \tag{2.7}$$

A recent discussion of the SLA and its variants can be found in Wood (2020).

2.3   INLA structure for time series

Let y^n denote the data (response vector), x^n a vector of latent Gaussian variables describing the model and $\boldsymbol{\theta}$ be a vector of hyperparameters. The dimension of the state vector x^n is large, typically being n in time series problems, whereas the dimension of the hyperparameters $\boldsymbol{\theta}$ is small, usually under 10. INLA uses a hierarchical framework to represent the underlying probabilistic structure, as discussed in the following two examples.

Gaussian Markov Random Field model

The original implementation of INLA assumes that that the latent vector x^n is defined by a Gaussian Markov Random Field (GMRF) (Rue and Held, 2005). The GMRF model can be expressed as a three level hierarchy consisting of the observed data y^n, the latent process x^n, and the unknown hyperparameters $\boldsymbol{\theta}$. We can represent the hierarchical structure as

$$y^n|x_t, \boldsymbol{\theta} \sim \prod_t p(y_t|\boldsymbol{x}_t, \boldsymbol{\theta}) \qquad \text{data model} \qquad (2.8)$$

$$\boldsymbol{x}^n|\boldsymbol{\theta} \sim N(\mathbf{0}, \boldsymbol{\Sigma}(\boldsymbol{\theta})) \qquad \text{GMRF prior} \qquad (2.9)$$

$$\boldsymbol{\theta} \sim \pi(\boldsymbol{\theta}) \qquad \text{hyperprior} \qquad (2.10)$$

where $x_\ell \perp x_m | \boldsymbol{x}^n{}_{(-\ell, m)}$ (see Chapter 2 – Appendix). The notation "$\boldsymbol{x}^n{}_{(-\ell, m)}$" indicates all the other elements of the parameter vector x^n excluding elements ℓ and m. The covariance matrix $\boldsymbol{\Sigma}$ depends on some hyperparameters $\boldsymbol{\theta}$.

Example: Local level model as a three level hierarchy

Consider the local level model described in (1.20) and (1.21). We can express this as a three level hierarchy with the observed data y^n, the latent unobserved process x^n, and the unknown random hyperparameters $\boldsymbol{\theta} = (\sigma_v^2, \sigma_w^2)'$ constituting the three levels of the hierarchy. The data distribution can belong to the exponential family (see Chapter 1 – Appendix) and could be normal, binomial, gamma, or Poisson, as we will see in later chapters, but the latent process is always Gaussian. The distributions of the hyperparameters need not be Gaussian. For instance, in this example, we may assume that the hyperparameters have the following distributions:

$$\frac{1}{\sigma_v^2} \sim \text{Gamma}(a_v, b_v) \text{ and } \frac{1}{\sigma_w^2} \sim \text{Gamma}(a_w, b_w),$$

where (a_v, b_v) and (a_w, b_w) are specified parameters for these prior distributions. We can represent the model in a hierarchical structure as

$$y^n|\boldsymbol{x}^n, \boldsymbol{\theta} \sim N(\boldsymbol{x}^n, \sigma_v^2 \boldsymbol{I}_n) \qquad \text{data model} \qquad (2.11)$$

$$\boldsymbol{x}^n|\boldsymbol{\theta} \sim \prod_{t=1}^{n} \pi(x_t|x_{t-1}, \boldsymbol{\theta}) \qquad \text{Markov prior} \qquad (2.12)$$

$$\boldsymbol{\theta} \sim \pi(\boldsymbol{\theta}) \qquad \text{hyperprior} \qquad (2.13)$$

where, $\pi(x_t|x_{t-1}, \boldsymbol{\theta})$ is $N(x_{t-1}, \sigma_w^2)$, $\boldsymbol{I}_n$ is the n-dimensional identity matrix, and $\pi(\boldsymbol{\theta})$ is given by the product of two gamma densities.

2.3.1 INLA steps

Using the hierarchical structure, INLA describes the posterior marginals of interest as

$$\pi(x_i|\boldsymbol{y}^n) = \int \pi(x_i|\boldsymbol{\theta}, \boldsymbol{y}^n)\pi(\boldsymbol{\theta}|\boldsymbol{y}^n)d\boldsymbol{\theta},$$

$$\pi(\theta_j|\boldsymbol{y}^n) = \int \pi(\boldsymbol{\theta}|\boldsymbol{y}^n)d\boldsymbol{\theta}_{(-j)}, \tag{2.14}$$

where θ_j denotes the jth component of $\boldsymbol{\theta}$, and $\boldsymbol{\theta}_{(-j)}$ denotes all the other components except the jth. Under the INLA approach, these forms are used to construct nested approximations

$$\tilde{\pi}(x_i|\boldsymbol{y}^n) = \int \tilde{\pi}(x_i|\boldsymbol{\theta}, \boldsymbol{y}^n)\tilde{\pi}(\boldsymbol{\theta}|\boldsymbol{y}^n)d\boldsymbol{\theta},$$

$$\tilde{\pi}(\theta_j|\boldsymbol{y}^n) = \int \tilde{\pi}(\boldsymbol{\theta}|\boldsymbol{y}^n)d\boldsymbol{\theta}_{(-j)}, \tag{2.15}$$

where $\tilde{\pi}(.|.)$ is an approximated conditional density. The INLA approach consists of three steps.

Step 1. First, INLA approximates the posterior marginal distributions of the hyperparameters $\pi(\boldsymbol{\theta}|\boldsymbol{y}^n)$ by using a Laplace approximation. To achieve this, the full conditional distribution of $\boldsymbol{x}^n$, $\pi(\boldsymbol{x}^n|\boldsymbol{\theta}, \boldsymbol{y}^n)$, is approximated using a multivariate Gaussian density $\tilde{\pi}_G(\boldsymbol{x}|\boldsymbol{\theta}, \boldsymbol{y}^n)$. Then, the posterior density of the hyperparameters is approximated by using the Laplace approximation

$$\tilde{\pi}(\boldsymbol{\theta}|\boldsymbol{y}^n) \propto \left. \frac{\pi(\boldsymbol{x}^n, \boldsymbol{\theta}, \boldsymbol{y}^n)}{\tilde{\pi}_G(\boldsymbol{x}^n|\boldsymbol{\theta}, \boldsymbol{y}^n)} \right|_{\boldsymbol{x}^n = \widehat{\boldsymbol{x}}^n}, \tag{2.16}$$

where $\widehat{\boldsymbol{x}}^n$ is the mode of the full conditional distribution $\pi(\boldsymbol{x}^n|\boldsymbol{\theta}, \boldsymbol{y}^n)$. This can be done using the Newton-Raphson algorithm.

Step 2. Second, INLA computes $\pi(x_i|\boldsymbol{\theta}, \boldsymbol{y}^n)$; i.e., it approximates conditional distributions of the latent Gaussian variables given selected hyperparameter values and the data, using one of three methods. In order of increasing accuracy, they are

a) the Gaussian approximation,

b) the simplified Laplace approximation, and

c) the Laplace approximation.

Step 3. The last step uses numerical integration, i.e.,

$$\tilde{\pi}(x_i|\boldsymbol{y}^n) = \sum_j \tilde{\pi}(x_i|\boldsymbol{\theta}_j, \boldsymbol{y}^n)\tilde{\pi}(\boldsymbol{\theta}_j|\boldsymbol{y})\Delta_j \tag{2.17}$$

to integrate out the hyperparameters, and obtains an approximation of the posterior distribution of the latent Gaussian variables.

Although in Step 2, the Laplace approximation is preferred in general, it can be computationally expensive. If the Gaussian approximation or the simplified Laplace approximation delivers small Kullback-Leibler divergence (see Chapter 2 - Appendix for a definition) between $\pi(x_i|\boldsymbol{y}^n)$ and $\tilde{\pi}(x_i|\boldsymbol{y}^n)$, then these methods are considered acceptable. Ruiz-Cárdenas et al. (2012) describe simulation studies and examples to illustrate fast and accurate Bayesian inference for dynamic models using the R-INLA package.

2.4   Forecasting in INLA

One of the main objectives of time series analysis is the prediction (or forecasting) of future values of Y_t, $t = n+1, n+2, \ldots$ given observed data $\boldsymbol{y}^n$ up to time period n. As discussed in Section 1.3, the predictive distribution of Y_{n+1} given $\boldsymbol{y}^n$ can be obtained, via the rules of probability, as

$$p(y_{n+1}|\boldsymbol{y}^n) = \int p(y_{n+1}|x_{n+1}, \boldsymbol{\theta}, \boldsymbol{y}^n)\pi(x_{n+1}, \boldsymbol{\theta}|\boldsymbol{y}^n)dx_{n+1}d\boldsymbol{\theta}. \tag{2.18}$$

Due to the conditional independence of Y_t's given x_t and $\boldsymbol{\theta}$, (2.18) reduces to

$$p(y_{n+1}|\boldsymbol{y}^n) = \int p(y_{n+1}|x_{n+1}, \boldsymbol{\theta})\pi(x_{n+1}, \boldsymbol{\theta}|\boldsymbol{y}^n)dx_{n+1}d\boldsymbol{\theta}, \tag{2.19}$$

which is referred to as the one-step ahead forecast distribution. Generalization of (2.18) to the k-step ahead forecast distribution for Y_{n+k} is given by

$$p(y_{n+k}|\boldsymbol{y}^n) = \int p(y_{n+k}|x_{n+k}, \boldsymbol{\theta})\pi(x_{n+k}, \boldsymbol{\theta}|\boldsymbol{y}^n)dx_{n+k}d\boldsymbol{\theta}, \tag{2.20}$$

for $k \geq 1$.

INLA provides us with approximations to the marginal posteriors $\pi(\boldsymbol{x}^n|\boldsymbol{y}^n)$ (marginalized over, or integrating out, the hyperparameter vector $\boldsymbol{\theta}$) and $\pi(\boldsymbol{\theta}|\boldsymbol{y}^n)$ (marginalized over the state vector $\boldsymbol{x}^n$), as well as samples from these marginal posterior distributions. To evaluate the one-step ahead predictive distribution in (2.19), we can write

$$\pi(x_{n+1}, \boldsymbol{\theta}|\boldsymbol{y}^n) = \pi(x_{n+1}|\boldsymbol{\theta}, \boldsymbol{y}^n)\pi(\boldsymbol{\theta}|\boldsymbol{y}^n). \tag{2.21}$$

Since samples from $\pi(\boldsymbol{\theta}|\boldsymbol{y}^n)$ are available, (2.19) can be evaluated by simulation, i.e., by drawing samples from $\pi(x_{n+1}|\boldsymbol{\theta}, \boldsymbol{y}^n)$. This is achieved in INLA by approximating the conditional posterior $\pi(x_{n+1}|\boldsymbol{\theta}, \boldsymbol{y}^n)$ and then drawing samples from the approximated distribution. For the case of a Gaussian observation equation as in (1.20), $\pi(x_{n+1}|\boldsymbol{\theta}, \boldsymbol{y}^n)$ can be approximated via a Gaussian distribution, but in general, more accurate approximations are used.[1] A similar strategy is implemented to obtain the k-step ahead forecast distribution (2.20).

It is important to note that INLA was developed as a general approach for Bayesian analysis using the GMRF structure for x_t's. As a result, it does not exploit the typical Markov evolution of x_t's in the dynamic model framework to simulate the forecast distributions.

2.5   Marginal likelihood computation in INLA

Following Rue et al. (2009), the marginal likelihood is computed as

$$m(\boldsymbol{y}^n|M_k) \approx \int \frac{p(\boldsymbol{y}^n, \boldsymbol{\theta}, \boldsymbol{x}^n|M_k)}{\tilde{\pi}_G(\boldsymbol{x}^n|\boldsymbol{y}^n, \boldsymbol{\theta}, M_k)}d\boldsymbol{\theta}, \tag{2.22}$$

[1]private communication with Håvard Rue (https://www.r-inla.org/contact-us)

where the right side is evaluated at $\boldsymbol{x}^n = \widehat{\boldsymbol{x}}^n(M_k)$, the mode of the conditional posterior density $\pi(\boldsymbol{x}^n|\boldsymbol{y}^n, \boldsymbol{\theta}, M_k)$ for model M_k. The cross validation marginal likelihood for a model M_k can also be approximated using INLA. The individual conditional predictive ordinate (CPO) for observation y_t is given by

$$p(y_t|\boldsymbol{y}_{(-t)}, M_k) = \int p(y_t|\boldsymbol{y}_{(-t)}, \boldsymbol{\theta}, M_k) \, \pi(\boldsymbol{\theta}|\boldsymbol{y}_{(-t)}, M_k) d\boldsymbol{\theta},$$

where, $\boldsymbol{y}_{(-t)}$ denotes all elements of $\boldsymbol{y}^n$ except y_t. As noted by Held et al. (2010), evaluation of the CPO requires $p(y_t|\boldsymbol{\theta}, M_k)$, which can be written as

$$p(y_t|\boldsymbol{y}_{(-t)}, \boldsymbol{\theta}, M_k) = 1 \left/ \int \frac{\pi(x_t|\boldsymbol{y}^n, \boldsymbol{\theta}, M_k)}{p(y_t|x_t, \boldsymbol{\theta}, M_k)} dx_t, \right.$$

where the numerator in the ratio, $\pi(x_t|\boldsymbol{y}^n, \boldsymbol{\theta}, M_k)$, is obtained using INLA and the denominator is simply the likelihood contribution of y_t. The integral is evaluated by numerical integration. Once we have the approximation $\tilde{p}(y_t|\boldsymbol{y}_{(-t)}, \boldsymbol{\theta}, M_k)$, the CPO term can be written as

$$\tilde{p}(y_t|\boldsymbol{y}_{(-t)}, M_k) = 1 \left/ \sum_j \frac{\tilde{\pi}(\boldsymbol{\theta}_j|\boldsymbol{y}^n, M_k)}{\tilde{p}(y_t|\boldsymbol{y}_{(-t)}, \boldsymbol{\theta}_j, M_k)} \Delta_j. \right. \tag{2.23}$$

The INLA estimator (2.23) is a *weighted harmonic mean* of the $\tilde{p}(y_t|\boldsymbol{y}_{(-t)}, \boldsymbol{\theta}_j, M_k)$ terms with weights $\tilde{\pi}(\boldsymbol{\theta}_j|\boldsymbol{y}^n, M_k)\Delta_j$. Given the individual CPO terms, we can obtain the cross validation marginal likelihood as their product.

2.6 R-INLA package – some basics

Integrated Nested Laplace Approximation (INLA) is implemented as an R package called INLA or R-INLA and is available from a specific repository at http://www.r-inla.org for Windows, Mac OS X, and Linux. The R-INLA website also includes extensive documentation about the package, examples, a discussion forum, and other resources about the theory and applications of INLA.

Note that the R-INLA package is not available from the Comprehensive R Archive Network (https://cran.r-project.org/) because it uses some external C libraries that make it difficult to build the binaries. Therefore, when installing the package, we need to use install.packages(), adding the URL of the R-INLA repository. A simple way to install the stable version is shown below. For the testing version, simply replace stable by testing when setting the repository.

```
install.packages(
  "INLA",
  repos = c(getOption("repos"),
            INLA = "https://inla.r-inla-download.org/R/stable"),
  dep = TRUE
)
```

To load the package in R once it is successfully installed, we need to type this in order to set up and run the code.

```
library(INLA)
```

The main function in the R-INLA package is called `inla()`. We will use this function together with `formula()` to set up and fit dynamic Bayesian models to time series using the INLA framework. In fitting a specified model to data, the `inla()` function is similar in usage to the R functions `lm()`, `glm()`, `arima()`, etc.

We can handle many different types of data/model combinations under the R-INLA umbrella. These include univariate time series with and without exogenous regressors, panel (multi-level) time series, multivariate time series, and discrete-valued time series such as binary or count time series. The modeling frameworks that we discuss in upcoming chapters include Gaussian linear models, hierarchical Gaussian linear models, generalized linear models with binomial, Poisson, or gamma sampling distributions, level correlated models for multivariate non-Gaussian time series, stochastic volatility models to handle time-varying heterogeneity, dynamic spatio-temporal models, etc. There are two main steps for fitting a model to data:

Step 1. Express the *linear predictor* part of the model as a `formula` object in R. This is similar to the expression on the right side in the familiar `lm(y~formula)` expression, and produces no output. Rather, `formula` becomes one of the inputs for the `inla()` call. We can specify *fixed effects* or *random effects* in the formulation.

- The fixed effect specification is similar to the way `lm()` or `glm()` includes such effects. The variables denoting the fixed effects are given in the formula, separated by "+". Fixed effects will have coefficients which are usually assigned vague priors, for instance, a $N(0, V^2)$ prior with large variance V^2. The prior parameters can be set through the option *control.fixed* in the `inla()` call. INLA allows two types of fixed effects, *linear* and *clinear*. These will be discussed in later chapters.

- A random effect is specified using the `f()` function, which includes an index to map the effect to the observations, specify the type of effect, etc. The effect is then included in the `formula`, where it is separated from other effects by a "+". Random effects will usually be assigned a common Gaussian prior with zero mean and an (unknown) precision, to which a prior (such as a gamma prior) may be assigned.

As we will see in examples throughout this book, the syntax of the `formula` consists of the response variable, followed by the ~ symbol, and then the fixed and/or random effects separated by + operators. Here is an example of a formula for a response y as a linear function of an intercept, two fixed effects predictors z1 and z2, and an additive i.i.d. error represented by id which is an index taking values $1 : n$.

```
formula <- y ~ z1 + z2 + f(id, model = "iid")
```

If we wish to exclude the intercept from the model, we modify the code as (recall that `lm()` uses this as well):

```
formula <- y ~ z1 + z2 + f(id, model = "iid") -1
```

or

```
formula <- y ~ 0 + z1 + z2 + f(id, model = "iid")
```

We include an i.i.d. random variable x corresponding to a random effect by modifying the code as follows:

```
formula <- y ~ 0 + z1 + z2 + f(x, model = "iid") + f(id, model = "iid")
```

Step 2. We run/fit the model by calling the `inla()` function with these input arguments:

a) a specified *formula* (described in Step 1);

b) a *family* for the data distribution; this is a string or a vector of strings to indicate the sampling distribution of the response (for constructing the likelihood function). Common families are Gaussian (default), Poisson, binomial, etc. A list of families is available by typing `names(inla.models()$likelihood`. Details about a specific family can be seen with `inla.doc("familyname")`;

c) the *data*, which consists of the data frame including the response, predictors, required mapping indexes, etc.;

d) a set of *control* options, such as

(i) *control.compute* to indicate which model selection criteria to compute,

(ii) *control.predictor* which gives details about the predictor and functions, such as a link function, or an indication about computation of posterior marginal distributions;

(iii) *control.family* which enables us to specify priors for parameters involved in the likelihood.

The `inla()` call returns an object that contains information from the fitted model that includes

1. Posterior summaries and marginals of parameters, hyperparameters, linear predictors, and fitted values, as well as a column *kld* representing the symmetric Kullback-Leibler divergence with the difference between the Gaussian and the simplified or full Laplace approximations for each posterior. R-INLA also provides a set of functions for post processing the posterior results.

2. Estimates of different criteria to assess and compare Bayesian models, such as marginal likelihood, conditional predictive ordinate (CPO), deviance information criterion (DIC), and the Watanabe-Akaike information criterion (WAIC).

Prior specification of parameters/hyperparamters is an important aspect of Bayesian modeling. R-INLA allows a user to assume default specifications for some items, while other items must be explicitly specified. We discuss prior specifications in Chapter 3.

It is worth noting that R-INLA does not have an explicit `predict()` function as we are used to seeing in `lm()` or `glm()`. We handle this by assigning NA's to responses that we wish to predict and including them with the observed data in the data frame.

INLA provides a few functions that operate on marginals. These functions and their usage are described in Chapter 14. We use `inla.tmarginal()` and `inla.zmarginal()` in Chapter 3, while other functions are used in later chapters as needed.

For `inla.tmarginal()`, the syntax is

```
inla.tmarginal(fun, marginal, n=1024, h.diff = .Machine$double.eps^(1/3),
               method = c("quantile", "linear"),  ...)
```

In the above chunk, the arguments as described in the `R-INLA` documentation are

- *marginal* is a marginal object from either `inla()` or `inla.hyperpar()`. This can either be *list(x=c(), y=c())* with density values y at locations x, or a *matrix(,n,2)* for which the density values are in the second column and the locations are in the first column. The `inla.hpdmarginal()` function assumes a unimodal density;

- *fun* is a (vectorized) function such as `function(x) exp(x)`.

The other arguments in `inla.tmarginal` may be left at default INLA specifications. The user only needs to specify the inverse transformation function through `fun`. For example, as we will discuss further in Chapter 3, suppose $\theta_v = \log(1/\sigma_v^2)$ is the internal INLA representation for a precision parameter $1/\sigma_v^2$, where σ_v^2 is a variance parameter. Then `inla.tmarginal` helps us recover information about the posterior marginal distribution of σ_v^2, via an inverse transformation from θ_v back to σ_v^2. This is done by using `fun`, which a user defines as $\exp(-\theta_v)$.

The syntax for `inla.zmarginal()` is

```
inla.zmarginal(marginal, silent = FALSE)
```

In most examples, we will need fitted values of the responses. When we run the `inla()` function, setting *compute = TRUE* in the *control.predictor* option will produce output which includes the following items:

summary.linear.predictor: is a data frame showing the posterior mean, standard deviation, and quantiles of the linear predictors;

summary.fitted.values: is a data frame with the posterior mean, standard deviation, and quantiles of the fitted values obtained by transforming the linear predictors by the inverse of the link function;

marginals.linear.predictor: is a list consisting of posterior marginals of the linear predictors;

marginals.fitted.values: is a list consisting of posterior marginals of the fitted values obtained by transforming the linear predictors via the inverse of the link function (in the Gaussian case, it is the identity link).

If we use the identity link function, as in Gaussian response models, the output from *summary.linear.predictor* and *summary.fitted.values* will coincide, and output from *marginals.linear.predictor* and *marginals.fitted.values* will coincide.

For more details, see Martino and Riebler (2014). Hubin and Storvik (2016) describe marginal likelihood computation, while Gómez-Rubio et al. (2020) describe Bayesian model averaging using `R-INLA`.

Chapter 2 – Appendix

The appendix contains important concepts and properties that are important to understand the theory and methods that underlie INLA.

Gaussian Markov Random Field (GMRF)

Let $\boldsymbol{x}^n = (x_1, \ldots, x_n)'$ have a normal distribution. Further, suppose it has Markov properties: $x_i \perp x_j | \boldsymbol{x}^n_{(-ij)}$, i.e., x_i and x_j are conditionally independent given all other components of $\boldsymbol{x}^n$, denoted by $\boldsymbol{x}^n_{(-ij)}$.

Consider an undirected graph $\mathcal{G} = \{\mathcal{V}, \mathcal{E}\}$ where $\mathcal{V} = \{1, 2, \ldots, n\}$ denotes the vertices and $\mathcal{E} = \{(i, j), i, j = 1, \ldots, n\}$ denotes the edges. If $x_i \perp x_j | \boldsymbol{x}^n_{(-ij)}$, then there is no edge between vertices i and j in the graph. On the other hand, there is an edge between i and j if x_i is not conditionally independent of x_j given $\boldsymbol{x}^n_{(-ij)}$.

Suppose the random vector $\boldsymbol{x}^n$ has mean $\boldsymbol{\mu}$ and precision matrix $\boldsymbol{\Omega}$, which is the inverse of the variance-covariance matrix of $\boldsymbol{x}^n$. Then, $\boldsymbol{x}^n$ is called a GRMF with respect to a graph $\mathcal{G}$ if and only if its p.d.f. has the form

$$\pi(\boldsymbol{x}^n) = (2\pi)^{-n/2} |\boldsymbol{\Omega}|^{1/2} \exp\left(-\frac{1}{2}(\boldsymbol{x}^n - \boldsymbol{\mu})' \boldsymbol{\Omega} (\boldsymbol{x}^n - \boldsymbol{\mu}) \right), \tag{2.24}$$

with $\Omega_{ij} \neq 0$ if and only if $(i, j) \in \mathcal{E}$, for all $i, j = 1, \ldots, n$.

Kullback-Leibler divergence

The Kullback-Leibler (KL) divergence is often referred to as relative entropy. It measures how one probability distribution differs from a reference probability distribution (Kullback and Leibler, 1951). If U and V are continuous-valued random variables with respective p.d.f.s $f(.)$ and $g(.)$, the (KL) divergence from $g(.)$ to $f(.)$ is defined as

$$D_{KL}(f \parallel g) = \int_{-\infty}^{\infty} f(u) \log \frac{f(u)}{g(u)} du. \tag{2.25}$$

The KL divergence is always non-negative, i.e., $D_{KL}(f \parallel g) \geq 0$ (Gibbs inequality), with equality holding if and only if $f(u) = g(u)$ almost everywhere.

If $f(.)$ represents a *true* distribution, we can find an approximation $g(.)$ by minimizing the KL-divergence (2.25). This is because $D_{KL}(f \parallel g)$ is the amount of information lost when $f(.)$ is approximated by $g(.)$. More details about the KL divergence can be found in Kullback (1959).

3

Details of R-INLA for Time Series

3.1 Introduction

In this chapter, we show how to use the R-INLA package for analyzing univariate time series. Specifically, we show how we set up the formula and model statements for analyzing a time series as a random walk, a random walk with drift, or autoregessions of various orders, excluding or including a level coefficient. R-INLA implements the integrated nested Laplace approximations that we discussed in Chapter 2 in order to carry out approximate Bayesian inference. We first show the details of Bayesian model fitting using R-INLA for time series simulated from a *random walk plus noise model* (also called the *local level model*), which was introduced in Chapter 1. We then discuss several other models for univariate time series. Further, we use these model fits to obtain future forecasts on hold-out (test) data. These forecasts can be used for predictive cross-validation and model selection. In practice, one can then use the best fitting model to forecast the time series into the future.

The first step is to install the package R-INLA:

```
install.packages(
  "INLA",
  repos = c(getOption("repos"),
            INLA = "https://inla.r-inla-download.org/R/stable"),
  dep = TRUE
)
```

Then, we source the package, along with other packages used in this chapter.

```
library(INLA)
library(tidyverse)
```

Next, the set of custom functions that we have built must be sourced so that they may be called at various places in the chapter.

```
source("functions_custom.R")
```

3.2 Random walk plus noise model

In Chapter 1, we defined the random walk plus noise model (also referred to as the local level model) using the observation equation and state equation shown in (1.20) and (1.21) as

$$y_t = x_t + v_t; \quad v_t \sim N(0, \sigma_v^2),$$
$$x_t = x_{t-1} + w_t; \quad w_t \sim N(0, \sigma_w^2),$$

where $x_0 = 0$, and v_t and w_t are uncorrelated errors, following zero mean normal distributions with unknown variances σ_v^2 and σ_w^2 respectively. The response variable is denoted by y_t. The *state variable* x_t is a latent Gaussian variable. Let $\boldsymbol{\theta} = (\sigma_v^2, \sigma_w^2)'$ denote the vector of unknown hyperparameters. This is a special case of the DLM shown in (1.22) and (1.23) with $\alpha = 0$, $F_t = 1$ and $G_t = 1$. We show the detailed steps for fitting this model to a simulated time series and verify that we are able to recover the true parameters that we used to generate the series. We set the length of the time series to be $n = 500$. The true parameter values for the observation error variance and state error variance are $\sigma_v^2 = 0.5$ and $\sigma_w^2 = 0.25$ respectively.

We simulate a time series from the model in (1.20) and (1.21) using the custom function `simulation.from.rw1()` which is included in the set of functions that we sourced in the introduction via

```
source("functions_custom.R")
```

To see a plot of the time series, use `plot.data = TRUE`. Figure 3.1 shows a wandering behavior which is typical of the nonstationary random walk series.

```
sim.rw1 <-
  simulation.from.rw1(
    sample.size = 500,
    burn.in = 100,
    level = 0,
    drift = 0,
    V = 0.5,
    W = 0.25,
    plot.data = TRUE,
    seed = 123457
  )
sim.rw1$sim.plot
```

To use `R-INLA` for fitting this model to the data, we discuss three main components:

(i)   a `formula` to specify the model, similar to the formula in the R function `lm()` used to fit the usual linear regression model (see Section 3.2.1).

(ii) model *execution*, which is used to obtain results and posterior summaries (see Section 3.2.2), and

(iii) *prior* and *hyperprior* specifications, when relevant (see Section 3.2.3).

3.2.1   R-INLA model formula

The information in the observation and state equations of a DLM is represented as a formula, similar to the formula used in the R function `lm()`. We can specify time effects, and in general any structured effect, using a function `f()` within the formula. Examples of structured effects for univariate time series are *iid, rw1, ar1*, etc. The time effect is represented as an index in `f()`. Specifically, we set up a time index *id.w* that runs from 1 to n, where n denotes the

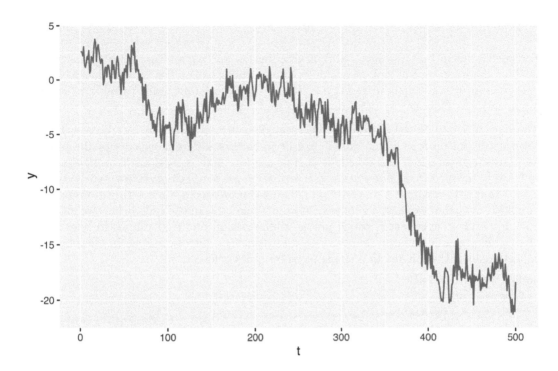

FIGURE 3.1: Simulated random walk plus noise series.

length of the time series. We then set up a data frame *rw1.dat*, with two columns containing the time series y_t and the time index *id.w*.

```
y.rw1 <- sim.rw1$sim.data
n <- length(y.rw1)
id.w <- 1:n
rw1.dat <- cbind.data.frame(y.rw1, id.w)
```

The *model* argument within the function f() is specified to be *rw1*. Since no intercept needs to be estimated in the observation equation (i.e., $\alpha = 0$), we include "-1" in the formula; note that this usage is similar to the lm() function.

```
formula.rw1 <- y.rw1 ~ f(id.w, model = "rw1",
                         constr = FALSE) - 1
```

In the formula, constr is an option to let the sum of terms (in this case x_t) be zero, which is the default option in R-INLA. Using ?f(), we can get help on the constr option: "*A boolean variable indicating whether to set a sum to 0 constraint on the term. By default the sum to 0 constraint is imposed on all intrinsic models ("iid","rw1","rw2","besag", etc.).*" We specify constr=FALSE, since there is no constraint in the random walk plus noise model. Information on all arguments available with the function f() can be accessed using ?f().

3.2.2 Model execution

We can fit the model using the function inla(). The general syntax is shown below, where formula was defined in the previous subsection.

```
inla(formula = ,family = ,data = ,control.compute = ,
   control.predictor = ,control.family = ,...)
```

For the Gaussian DLM, y_t has a normal distribution and we set `family = gaussian`. In later chapters, we will look at examples of non-Gaussian families for y_t, and specify these via the appropriate input in `family =`. Recall that the latent state variable x_t is always assumed to be Gaussian in this framework (see Chapter 2).

The data is set to be the data frame we created earlier, i.e., `data = rw1.dat`. R-INLA has several control options which enable a user to control different aspects of the modeling and output generation, and excellent documentation of these options is available on their website by using `?inla`. For example, information on `control.compute` can be obtained from `?control.compute`. Gómez-Rubio (2020) has presented a table with various options in `control.compute`; we show a similar display in Table 14.2. In the code below, we include the option `control.predictor`, which provides information on fitted values. In later sections of this chapter and in later chapters, we discuss other control options. In Chapter 14, we collect some of these items in a single place for quick reference.

```
model.rw1 <- inla(
   formula.rw1,
   family = "gaussian",
   data = rw1.dat,
   control.predictor = list(compute = TRUE)
)
```

The results from the `inla()` call shown above are stored in *model.rw1*, which is a list consisting of a number of objects. Before we do a deep dive into the output generated by `inla()`, we look at the detailed model summary, which is obtained using the function `summary()`:

```
summary(model.rw1)
##
## Call:
##    c("inla(formula = formula.rw1, family = \"gaussian\", data =
##    rw1.dat, ", " control.predictor = list(compute = TRUE))")
## Time used:
##    Pre = 4.14, Running = 0.572, Post = 0.422, Total = 5.13
## Random effects:
##   Name      Model
##      id.w RW1 model
##
## Model hyperparameters:
##                                           mean    sd 0.025quant
## Precision for the Gaussian observations 1.66 0.152       1.38
## Precision for id.w                      4.79 0.868       3.32
##                                         0.5quant 0.975quant mode
## Precision for the Gaussian observations     1.65       1.98 1.64
## Precision for id.w                          4.71       6.72 4.56
##
## Expected number of effective parameters(stdev): 143.47(15.72)
## Number of equivalent replicates : 3.48
##
```

```
## Marginal log-Likelihood:   -750.25
## Posterior marginals for the linear predictor and
##   the fitted values are computed
```

Description of the output

The output from **summary()** is very detailed, parts of which are self-explanatory, while others need some explanation. Below, we describe all the terms from the output.

1. *call*: the function called, with relevant arguments. Here,

```
summary(model.rw1)$call
## [1] "inla(formula = formula.rw1, family = \"gaussian\", data = rw1.dat, "
## [2] "    control.predictor = list(compute = TRUE))"
```

We can also obtain this using **model.rw1$call**.

2. *Time used*: time taken by the system to execute the model fit. Here,

```
summary(model.rw1)$cpu.used
## [1] "Pre = 4.14, Running = 0.572, Post = 0.422, Total = 5.13"
```

We can also obtain the same output by using **model.rw1$cpu.used**.

3. *The model has no fixed effects*: note that in the equations for the random walk plus noise model, x_t is time varying, and there are no "fixed" effects in the model (such as the intercept, α, say).

4. *Random effects*: in the model, x_t is a latent process and the structured effect *rw1* is captured through the variable *id.w*. See

```
summary(model.rw1)$random.names
## [1] "id.w"
summary(model.rw1)$random.model
## [1] "RW1 model"
```

5. *Model hyperparameters*: a summary of parameter estimation corresponding to σ_v^2 and σ_w^2 is shown under this heading, and consists of the mean, standard deviation, 2.5^{th} quantile, median, 97.5^{th} quantile, and the mode of the posterior distributions of the following terms: precision for the Gaussian observations, i.e., the precision for v_t, which is $1/\sigma_v^2$, and precision for *id.w*, i.e., the precision of w_t, which is $1/\sigma_w^2$.

```
summary(model.rw1)$hyperpar
##                                        mean     sd 0.025quant
## Precision for the Gaussian observations 1.661 0.152      1.380
## Precision for id.w                      4.791 0.868      3.316
##                                        0.5quant 0.975quant  mode
## Precision for the Gaussian observations    1.654      1.978 1.643
## Precision for id.w                         4.712      6.720 4.559
```

We can also get this using `model.rw1$summary.hyperpar`; note the difference in the number of digits printed.

6. *Expected number of effective parameters*: this is the number of *independently estimated* parameters in the model.

```
summary(model.rw1)$neffp
##                                          [,1]
## Expectected  number of parameters 143.475
## Stdev of the number of parameters   15.720
## Number of equivalent replicates      3.485
```

7. *Marginal log-likelihood*: this was defined in Chapter 2 and is used in model selection, as discussed in Section 3.8.

```
summary(model.rw1)$mlik
##                                              [,1]
## log marginal-likelihood (integration) -750.0
## log marginal-likelihood (Gaussian)    -750.3
```

We can also obtain this using `model.rw1$mlik`.

8. *Posterior marginals for linear predictor and fitted values*: while calling the `inla()` function, we request this output through the option `control.predictor = list(compute = TRUE)`.

```
head(summary(model.rw1)$linear.predictor)
```

This can also be obtained using `model.rw1$summary.linear.predictor`. It is useful to print a condensed summary for the fitted values:

```
head(summary(model.rw1)$linear.predictor)
##                   mean    sd 0.025quant 0.5quant 0.975quant  mode kld
## Predictor.001 2.488 0.518      1.473    2.488      3.504 2.487   0
## Predictor.002 2.428 0.450      1.547    2.428      3.310 2.428   0
## Predictor.003 2.331 0.426      1.496    2.331      3.167 2.330   0
## Predictor.004 2.270 0.419      1.449    2.269      3.093 2.268   0
## Predictor.005 1.918 0.417      1.099    1.918      2.734 1.919   0
## Predictor.006 1.745 0.417      0.925    1.746      2.561 1.747   0
```

Condensed output

The output from `inla()` can be unwieldy. It is possible to show a condensed output by excluding selected columns from the detailed output, as we show below. Our function `format.inla.out()` requires the R package `tidyverse` (Wickham et al., 2019). For details on the `tidyverse` package and other data wrangling approaches, see Boehmke (2016). To show only the mean, standard deviation, and median of the hyperparameter distributions, we use

```
format.inla.out(model.rw1$summary.hyperpar[,c(1,2,4)])

##   name                     mean   sd    0.5q
## 1 Precision for Gaussian obs 1.661 0.152 1.654
```

```
## 2 Precision for id.w          4.791 0.868 4.712
```

The condensed output below only shows the three quantiles from the hyperparameter distributions.

```
format.inla.out(model.rw1$summary.hyperpar[,c(3,4,5)])
```

```
  name                       0.025q  0.5q   0.975q
1 Precision for Gaussian obs 1.380   1.654  1.978
2 Precision for id.w         3.316   4.712  6.720
```

Below, we illustrate output only on the mean, median, and mode of the hyperparameter distributions.

```
format.inla.out(model.rw1$summary.hyperpar[,c(1,4,6)])
```

```
  name                       mean   0.5q   mode
1 Precision for Gaussian obs 1.661  1.654  1.643
2 Precision for id.w         4.791  4.712  4.559
```

Although we only show condensed output in most places in the book due to space restrictions, a user may obtain and directly view the full output by excluding `format.inla.out()`.

3.2.3 Prior specifications for hyperparameters

Setting prior specifications for the hyperparameters is an important step in the Bayesian framework. Data analysts often give considerable thought to specifying appropriate and reasonable priors. If a user does not wish to specify a custom prior distribution, R-INLA uses a default prior specification. In most situations, these default priors work quite well. Here, we talk about *default priors* available for hyperparameters (or model parameters) in a Gaussian DLM. In Section 3.9, we discuss how a user may set up non-default or custom priors (also see Chapter 5 in Gómez-Rubio (2020)).

It is important to note that R-INLA assigns priors via an *internal representation* for the parameters. This representation may not be the same as the scale of the parameter that is specified in the model by a user. For instance, precisions (which are reciprocals of variances) are represented on the *log scale*. In the Gausssian DLM, the precision corresponding to the observation noise variance is $1/\sigma_v^2$, and R-INLA sets a prior for $\theta_v = \log(1/\sigma_v^2)$. As explained in Rue et al. (2009), the internal representation is computationally convenient, since the parameter is not bounded in the internal scale and has the entire real line $\mathbb{R}$ as support. We next show how we set up and execute the model under default priors for observation and state noise variances via $\theta_v = \log(1/\sigma_v^2)$ and $\theta_w = \log(1/\sigma_w^2)$.

Default priors for σ_v^2 and σ_w^2

To define the prior specification corresponding to the observation noise variance σ_v^2, consider the corresponding log precision

$$\theta_v = \log\left(\frac{1}{\sigma_v^2}\right),$$

and let

$$\theta_v \sim \log \text{gamma}(\text{shape} = 1, \text{inverse scale} = 5e-05).$$

The logarithm of the precision θ_v is assumed by default to follow a log-gamma distribution. The shape and inverse scale parameters of this log-gamma prior are unknown, and by default, R-INLA assigns to them the values 1 and 5e-05 respectively. The default initial value for this log-gamma hyperparameter is set as 4.

Similarly, a default log-gamma prior with values 1 and 5e-05 for the shape and inverse scale and a default initial value of 4 is specified for $\theta_w = \log(1/\sigma_w^2)$, the logarithm of the reciprocal of the state error variance σ_w^2. That is, we let

$$\theta_w = \log\left(\frac{1}{\sigma_w^2}\right),$$

and

$$\theta_w \sim \log \text{gamma}(\text{shape} = 1, \text{inverse scale} = 5e - 05).$$

More details on these default prior specifications for this model can be accessed from the INLA documentation `inla.doc("rw1")`.

3.2.4   Posterior distributions of hyperparameters

We can obtain marginal posterior distributions of the latent variables (treated as parameters) x_t and the hyperparameters $\boldsymbol{\theta}$, and manipulate these distributions. The options are summarized in Table 14.3, which shows the different functions and their descriptions (similar to the one displayed in Gómez-Rubio (2020)).

We have seen that R-INLA works with the log precisions of hyperparameters rather than their variances. Therefore, in order to obtain marginal posterior distributions of σ_v^2 and σ_w^2, we must transform each log(precision) back to the corresponding variance using the function `inla.tmarginal()`. The result for σ_v^2 is shown in Figure 3.2.

```
sigma2.v.dist <- inla.tmarginal(
  fun = function(x)
    exp(-x),
  marginal = model.rw1$
  internal.marginals.hyperpar$`Log precision for the Gaussian observations`
)
plot(
  sigma2.v.dist,
  type = "l",
  xlab = expression(paste(
    "Observation noise variance ", sigma[v] ^ 2, sep = " "
  )),
  ylab = "density"
)
```

The marginal posterior for σ_w^2 may be similarly obtained, see Figure 3.3.

```
sigma2.w.dist <- inla.tmarginal(
  fun = function(x)
    exp(-x),
  marginal = model.rw1$internal.marginals.hyperpar$`Log precision for id`
)
```

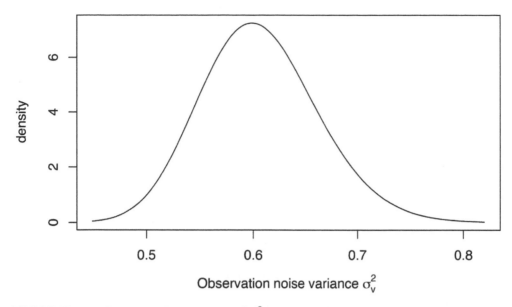

FIGURE 3.2: Posterior distribution of σ_v^2 in the random walk plus noise model.

```
plot(sigma2.w.dist,
    type = "l",
     xlab = expression(paste(
   "State noise variance ", sigma[w]^2, sep = " "
 )),
    ylab = "density")
```

R-INLA incorporates several functions to manipulate the posterior marginals. For example, `inla.emarginal()` and `inla.qmarginal()` respectively calculate the expectation and quantiles of the marginal posteriors. The function `inla.smarginal()` can be used to do spline smoothing, `inla.tmarginal()` can be used to transform the marginals, and `inla.zmarginal()` provides summary statistics.

We can write the posterior expectations of σ_v^2 and σ_w^2.

```
sigma2.v.hat <-
  inla.emarginal(
    fun = function(x)
      exp(-x),
    marginal = model.rw1$internal.marginals.hyperpar$
      `Log precision for the Gaussian observations`
  )
sigma2.w.hat <-
  inla.emarginal(
    fun = function(x)
      exp(-x),
```

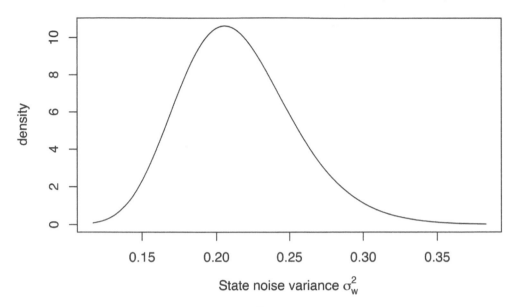

FIGURE 3.3: Posterior distribution of σ_w^2 in the random walk plus noise model.

```
    marginal = model.rw1$internal.marginals.hyperpar$`Log precision for id`
  )
cat(paste(
  "Estimated observation noise variance, sigma2.v",
  round(sigma2.v.hat, 2),
  sep = " = "
),
"\n")
## Estimated observation noise variance, sigma2.v = 0.61
cat(paste(
  "Estimated state noise variance, sigma2.w",
  round(sigma2.w.hat, 2),
  sep = " = "
))
## Estimated state noise variance, sigma2.w = 0.22
```

3.2.5 Fitted values for latent states and responses

We describe how we recover the fitted state variables x_t and the fitted responses y_t for each time $t = 1, \ldots, n$. The fitted state variable from R-INLA corresponds to the smoothed estimate of x_t, as we discuss below.

Fitted values for the state variables x_t

For $t = 1, \ldots, n$, fitted values of the state variable x_t are obtained as the means of the posterior distributions of x_t given all the data $\boldsymbol{y}^n = (y_1, \ldots, y_n)'$, i.e., $\pi(x_t | \boldsymbol{y}^n)$. Posterior summaries of the state variable (the local level variable in the case of the random walk plus noise model) x_t are stored in *summary.random*. We display the first few rows below.

```
head(model.rw1$summary.random$id.w)
##    ID   mean      sd 0.025quant 0.5quant 0.975quant   mode        kld
## 1  1 2.488 0.5178     1.4718    2.488      3.505 2.487 8.692e-08
## 2  2 2.428 0.4497     1.5453    2.428      3.311 2.428 1.188e-07
## 3  3 2.331 0.4262     1.4943    2.331      3.168 2.330 9.244e-08
## 4  4 2.270 0.4191     1.4475    2.269      3.093 2.268 1.318e-07
## 5  5 1.918 0.4166     1.0983    1.918      2.734 1.919 8.678e-08
## 6  6 1.745 0.4170     0.9234    1.746      2.561 1.747 2.460e-07
```

A condensed output is shown below.

```
format.inla.out(head(model.rw1$summary.random$id[,c(1:5)]))
##    ID mean    sd      0.025q 0.5q
## 1  1   2.488 0.518 1.472    2.488
## 2  2   2.428 0.450 1.545    2.428
## 3  3   2.331 0.426 1.494    2.331
## 4  4   2.270 0.419 1.448    2.269
## 5  5   1.918 0.417 1.098    1.918
## 6  6   1.745 0.417 0.923    1.746
```

In the usual Gaussian DLM terminology, these estimates are called the Kalman smoothed estimates or simply, the *smooth* (Shumway and Stoffer, 2017), and are obtained for $t = 1, \ldots, n$ using just one `inla()` call. By default, R-INLA uses a *simplified Laplace approximation* (see Chapter 2 for a description). For each t, posterior summaries of $\pi(x_t | \boldsymbol{y}^n)$ are given, along with the Kullback-Leibler Divergence (KLD) defined in (2.25). The KLD gives the distance between the standard Gaussian distribution and the *simplified Laplace approximation* to the marginal posterior densities (see Chapter 2 – Appendix for a description).

A plot of the posterior means of x_t over time together with the 2.5th and 97.5th percentiles for the random walk plus noise model is shown in Figure 3.4.

```
plot.data <- model.rw1$summary.random$id %>%
   select(time = ID, mean,
          "0.025quant", "0.975quant")
multiline.plot(plot.data, xlab = "t", ylab = "posterior mean",
               line.color = c("black", "red", "red"),
               line.size = 0.6)
```

For plotting multiple lines, we have used a customized function called `multiline.plot()`, which uses the R package `ggplot2` (Wickham, 2016). A summary of this function and other customized functions is shown in Chapter 14.

Fitted values for y_t

For $t = 1, \ldots, n$, the fitted values of y_t are the posterior means of the predictive distributions of y_t given $\boldsymbol{y}^n = (y_1, \ldots, y_n)'$ and can be retrieved by using the code below; see Figure 3.5 for a plot. As we discussed in Chapter 2, for Gaussian response models (where the link is the identity

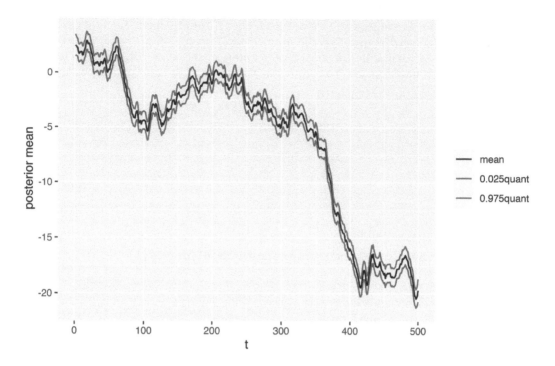

FIGURE 3.4: Posterior means and percentiles of the state variable over time in the random walk plus noise model.

link), we can use either the `summary.linear.predictor` or the `summary.fitted.values` in the `control.predictor` option in `inla()`, setting `compute = TRUE`.

```
fit <- model.rw1$summary.linear.predictor$mean
df <- as_tibble(cbind.data.frame(time = 1:n, y.rw1, fit))
multiline.plot(
  plot.data = df,
  title = "",
  xlab = "t",
  ylab = "y",
  line.type = "dashed",
  line.color = c("red", "blue"),
  line.size = 0.6
)
```

Two other approaches to obtain the fitted values are shown below. For Gaussian DLMs, each of these three ways gives exactly the same results. Other summaries such as standard deviation (sd) or quantiles can be obtained similarly. However, for non-Gaussian DLMs, note that we cannot use `summary.linear.predictor`, but can instead use one of the two alternate approaches shown below. We will revisit this in Chapter 9 for count time series. See `?inla` for more details.

```
fit.alt.1 <- model.rw1$summary.fitted.values$mean
fit.alt.2 <- model.rw1$marginals.linear.predictor %>%
  sapply(function(x) inla.emarginal(fun = function(y) y, marginal =x))
```

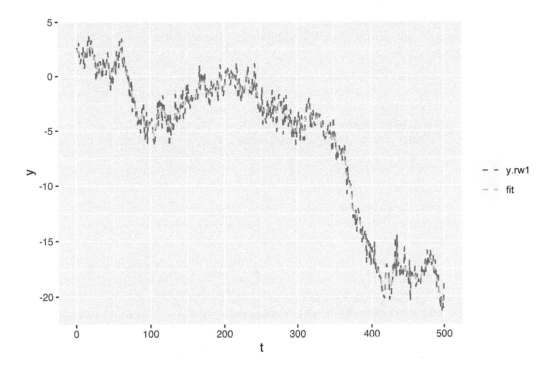

FIGURE 3.5: Simulated and fitted responses under the random walk plus noise model.

3.2.6 Filtering and smoothing in DLM

As we mentioned in Chapter 1, the Kalman filtering algorithm is a recursive algorithm to estimate the state variable x_t in a DLM given data/information up to time t, i.e., given y^t (Shumway and Stoffer, 2017). Using this algorithm, a time series analyst is able to compute filter estimates of x_t for $t = 1, \ldots, n$. Kalman filtering thus enables *recursive estimation* of the state variable as information comes in sequentially over time. It is well known that Kalman smoothing, on the other hand, estimates x_t for times $t = 1, \ldots, n$ given *all* the data y^n. When we use R-INLA to fit a DLM, it returns the *smoothed* estimates of x_t, $t = 1, \ldots, n$. The posteriors of interest are therefore also approximated using all the information y^n. The forward recursion to obtain filter estimates and backward recursions to obtain the smoothed estimates underlies Kalman's algorithm (Kalman (1960)). These estimates are then used to recursively estimate the states and the hyperparameters. As discussed in Ruiz-Cárdenas et al. (2012), a recursive (or sequential) approach for estimation and prediction is necessary in *online* (or real-time) settings as each new observation arrives, for instance in computer vision, finance, IoT monitoring systems, real-time traffic monitoring, etc. In these examples, it is useful to obtain *filter estimates* of the state variable x_t, and then obtain the corresponding fits for y_t.

The INLA approach argues that the estimation need not be recursive (or *dynamic*) in situations where all n observations in the time series are available rather than trickling in sequentially. The posteriors of interest are directly approximated in a non-sequential manner. While filter estimates of the state variable cannot be directly output in R-INLA, we can obtain filter estimates for x_t, $t = 1, \ldots, n$ by repeatedly calling the `inla()` function n times. This is clearly a computationally expensive exercise, and when possible, it is useful to run the n `inla()` calls in parallel. We implement the parallel runs using the R package `doParallel()`.

We obtain filter estimates of x_t from the random walk plus noise model, using the series y_t that we simulated earlier in this section; see Figure 3.6. The custom function to obtain the filter estimates, `filt.inla()`, is shown in Section 14.3 and is included in the set of functions we sourced at the beginning of the chapter. Compared with the smoothed estimates for x_t, computation of the filter estimates is time-consuming as expected (since we call `inla()` n times).

```
out <-
  filt.inla(
    data.series = y.rw1,
    model = "rw1",
    trend = "no",
    alpha = "no"
  )

filt.all.rw1 <- out$filt.all.bind
filt.estimate.rw1 <- c(NA, filt.all.rw1$mean)
smooth.est <- model.rw1$summary.random$id$mean
plot.df <- as_tibble(cbind.data.frame(time = id.w,
                                      y = y.rw1,
                                      filter = filt.estimate.rw1,
                                      smooth = smooth.est))
multiline.plot(plot.df,
               title = "",
               xlab = "t",
               ylab = " ", line.type = "solid", line.size = 0.6,
               line.color = c("red", "yellow", "green"))
```

3.3   AR(1) with level plus noise model

In Chapter 1, we defined the AR(1) plus noise model using the observation and state equations (1.24) and (1.25). Here, we describe the R-INLA fit for an AR(1) with level plus noise model defined by the following DLM equations:

$$y_t = \alpha + x_t + v_t; \quad v_t \sim N(0, \sigma_v^2), \tag{3.1}$$

$$x_t = \phi x_{t-1} + w_t; \quad w_t \sim N(0, \sigma_w^2), \ t = 2, \ldots, n, \tag{3.2}$$

and

$$x_1 \sim N(0, \sigma_w^2/(1 - \phi^2)), \tag{3.3}$$

where $|\phi| < 1$ and α is the level parameter. Let $\tau = 1/\sigma_w^2$ and $\kappa = \tau(1 - \phi^2) = (1 - \phi^2)/\sigma_w^2$ denote the marginal precision of the state x_t.

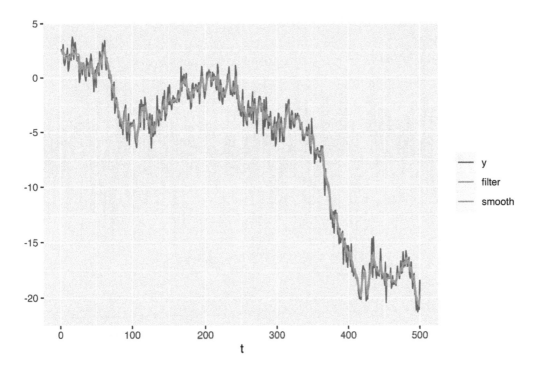

FIGURE 3.6: Simulated responses (red), with filtered (yellow) and smoothed (green) estimates of the state variable in the random walk plus noise model.

Note: R-INLA uses the notation ρ instead of ϕ in the AR(1) model. In the output, when we see *Rho*, it refers to ϕ in our notation.

We use the custom function `simulation.from.ar()` to simulate $n = 500$ values x_t from a Gaussian AR(1) process with $\sigma_w^2 = 0.1$ and $\phi = 0.6$. This is done using the R function `arima.sim()`. We add $\alpha = 1.4$ and Gaussian errors v_t with variance $\sigma_v^2 = 0.2$ to x_t in order to get the process y_t; see Figure 3.7.

```
sim.ar1.level <-
  simulation.from.ar(
    sample.size = 500,
    burn.in = 100,
    phi = 0.6,
    level = 1.4,
    drift = 0,
    V = 0.2,
    W = 0.1,
    plot.data = TRUE,
    seed = 123457
  )
sim.ar1.level$sim.plot
```

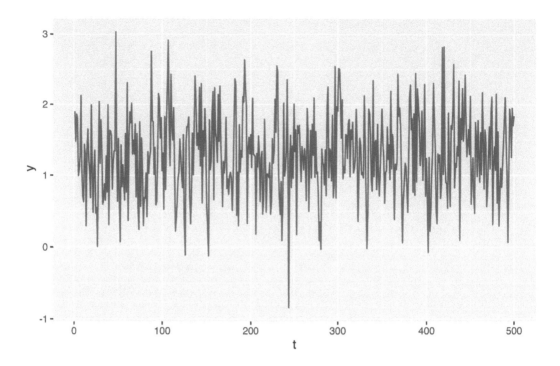

FIGURE 3.7: Simulated AR(1) with level plus noise series.

We next describe hyperparameter prior specifications. Let

$$\theta_v = \log(1/\sigma_v^2), \quad \theta_w = \log \kappa = \log \left(\frac{1 - \phi^2}{\sigma_w^2} \right),$$

$$\theta_\phi = \log \frac{1 + \phi}{1 - \phi}, \tag{3.4}$$

and let $\boldsymbol{\theta} = (\theta_v, \theta_w, \theta_\phi)'$ be the internal representation of the hyperparameters. The default priors on θ_v and θ_w are log-gamma priors with shape and inverse scale being 1 and 5e-05, and default initial values of 4 in each case. Assume that

$$\theta_\phi \sim N(\text{mean} = 0, \text{precision} = 0.15);$$

the normal prior depends on the mean and precision (reciprocal of the variance) parameters. R-INLA sets default values of 0 and 0.15 respectively for these values, while the initial value of θ_ϕ is assumed to be 2. For more details on the default prior specifications, see the R-INLA documentation `inla.doc("ar1")` and Chapter 5 of Gómez-Rubio (2020). The prior for the level parameter (intercept) α is assumed to be a normal distribution with mean 0 and precision 0.001. For more details on this, see the R-INLA documentation for `?control.fixed`. The code for model fitting is similar to the *rw1* example, except that we now give a formula for an AR(1) model by inserting `formula.ar1.level` as the first item in the call to `inla()`:

```
y.ar1.level <- sim.ar1.level$sim.data
n <- length(y.ar1.level)
id.x <- 1:n
```

```
data.ar1.level <- cbind.data.frame(y.ar1.level, id.x)
formula.ar1.level <-
  y.ar1.level ~ f(id.x, model = "ar1", constr = FALSE)
model.ar1.level <- inla(
  formula.ar1.level,
  family = "gaussian",
  data = data.ar1.level,
  control.predictor = list(compute = TRUE)
)
# summary(model.ar1.level)
format.inla.out(model.ar1.level$summary.hyperpar[,c(1:2)])
##    name                        mean   sd
## 1 Precision for Gaussian obs  4.509 0.591
## 2 Precision for id.x          8.170 1.853
## 3 Rho for id.x                0.450 0.075
format.inla.out(model.ar1.level$summary.hyperpar[,c(3:5)])
##    name                        0.025q 0.5q  0.975q
## 1 Precision for Gaussian obs  3.389  4.496  5.748
## 2 Precision for id.x          5.277  7.898 12.634
## 3 Rho for id.x                0.301  0.450  0.597
```

We can then explore the posterior estimates of σ_v^2, $Var(x_t) = \sigma_x^2$, ϕ, and α, using code similar to that in the *rw1* example. Specifically, for the AR(1) coefficient ϕ, the inverse transformation from θ_ϕ to ϕ is obtained as

$$\phi = \frac{2\exp(\theta_\phi)}{1 + \exp(\theta_\phi)} - 1. \tag{3.5}$$

For an AR(1) model, note that R-INLA returns the precision of x_t i.e., $1/Var(x_t)$. This is different from the output we saw earlier for the *rw1* model, where inla() returns the precision of the state error, $1/\sigma_w^2$. For the variance of the state variable, $Var(x_t)$, we have

```
var.x.dist <- inla.tmarginal(
  fun = function(x)
    exp(-x),
  marginal =
    model.ar1.level$internal.marginals.hyperpar$`Log precision for id`
)
var.x.zm <- inla.zmarginal(var.x.dist)
## Mean              0.128389
## Stdev             0.027596
## Quantile  0.025  0.0799905
## Quantile  0.25   0.10859
## Quantile  0.5    0.126634
## Quantile  0.75   0.145925
## Quantile  0.975  0.187606
```

The level (or intercept) is a fixed effect in this model. We can recover a summary of the fixed effect using

```
format.inla.out(model.ar1.level$summary.fixed[,c(1:5)])
```

```
##    name        mean   sd     0.025q  0.5q   0.975q
## 1 Intercept  1.287  0.034  1.221   1.287  1.354
```

The *Highest Posterior Density* (HPD) interval (Gamerman and Lopes, 2006) for α is

```
inla.hpdmarginal(0.95, model.ar1.level$marginals.fixed$`(Intercept)`)
##              low    high
## level:0.95 1.221  1.353
```

An alternate way to obtain posterior summaries for the fixed effect is given below.

```
alpha.fe <-  inla.tmarginal(
  fun = function(x)
    x,
  marginal = model.ar1.level$marginals.fixed$`(Intercept)`
)
alpha.fe.zm <- inla.zmarginal(alpha.fe, silent = TRUE)
alpha.hpd.interval <- inla.hpdmarginal(p = 0.95, alpha.fe)
```

The fitted values for x_t and y_t for the AR(1) with level plus noise model can be obtained using a code similar to that for the *rw1* model. This is available on the GitHub link associated with the book.

It is useful to illustrate the effect of ignoring the level α in fitting the model. While the correct (simulated) model is (3.1) and (3.2), suppose instead that a user fits a model without level by using the formula shown below.

```
formula.ar1.nolevel <-
  y.ar1.level ~ -1 + f(id.x, model = "ar1", constr = FALSE)
model.ar1.nolevel <- inla(
  formula.ar1.nolevel,
  family = "gaussian",
  data = data.ar1.level,
  control.predictor = list(compute = TRUE)
)
# summary(model.ar1.nolevel)
format.inla.out(model.ar1.nolevel$summary.hyperpar[,c(1:2)])
##    name                        mean    sd
## 1 Precision for Gaussian obs 2.843  0.181
## 2 Precision for id.x         3.100  1.791
## 3 Rho for id.x               1.000  0.000
format.inla.out(model.ar1.nolevel$summary.hyperpar[,c(3:5)])
##    name                        0.025q  0.5q   0.975q
## 1 Precision for Gaussian obs 2.504  2.837  3.217
## 2 Precision for id.x         0.844  2.715  7.642
## 3 Rho for id.x               0.998  1.000  1.000
```

Note that as a consequence of omitting α in the modeling, the estimate of the coefficient ϕ tends to 1, which would indicate a latent random walk dependence! To avoid this, we recommend that a user always fits a model which includes a level α in the observation equation. Of course, this coefficient will be estimated close to 0 when the level in the time series is zero.

3.4   Dynamic linear models with higher order AR lags

In R-INLA, we can assume that the latent Gaussian process x_t is an autoregressive process of order p (i.e., an AR(p) process) for $p \geq 1$, as described below.

AR(p) with level plus noise model

Assume that the latent process x_t follows a stationary Gaussian AR(p) process

$$\phi(B)x_t = (1 - \phi_1 B - \phi_2 B^2 - \ldots - \phi_p B^p)x_t = w_t, \tag{3.6}$$

where $w_t \sim N(0, \sigma_w^2)$, and the AR coefficients $\phi_1, \phi_2, \ldots, \phi_p$ satisfy the stationarity condition, i.e., the modulus of each root of the AR polynomial equation $\phi(B) = 0$ is greater than 1. See Chapter 3 – Appendix for more details, such as a definition of the backshift operator B, and checking the stationarity conditions using the `polyroot()` function in R.

The AR(p) with level plus noise model for y_t is defined as follows:

$$y_t = \alpha + x_t + v_t; \quad v_t \sim N(0, \sigma_v^2), \tag{3.7}$$

$$x_t = \sum_{j=1}^{p} \phi_j x_{t-j} + w_t; \quad w_t \sim N(0, \sigma_w^2), \tag{3.8}$$

where, α is the level; we assume that $x_0 = x_{-1} = \ldots = x_{1-p} = 0$, and the errors v_t and w_t follow zero mean normal distributions with unknown variances σ_v^2 and σ_w^2 respectively. R-INLA parameterizes the AR(p) process in terms of the partial autocorrelations (see Chapter 3 – Appendix). When $p = 2$,

$$r_1 = \phi_1/(1 - \phi_2), \text{ and } r_2 = \phi_2, \tag{3.9}$$

R-INLA estimates these together with the error precisions $1/\sigma_v^2$ and $1/\sigma_w^2$. Using the inverse of this 1-1 transformation (see (3.41) in Chapter 3 - Appendix), we can recover the estimates of $\boldsymbol{\phi} = (\phi_1, \ldots, \phi_p)'$. These steps are shown below for an AR(2) model.

Example: AR(2) with level plus noise model

We simulate an AR(2) with level plus noise model with true values $\phi_1 = 1.5$ and $\phi_2 = -0.75$ (corresponding to a stationary process), level $\alpha = 10$, and error variances $\sigma_v^2 = 1.25$ and $\sigma_w^2 = 0.05$. The plot of simulated series is shown in Figure 3.8.

```
sim.ar2 <-
  simulation.from.ar(
    sample.size = 500,
    burn.in = 100,
    phi = c(1.5,-0.75),
    level = 10,
    drift = 0,
    V = 1.25,
    W = 0.05,
    plot.data = TRUE,
    seed = 123457
```

```
  )
sim.ar2$sim.plot
```

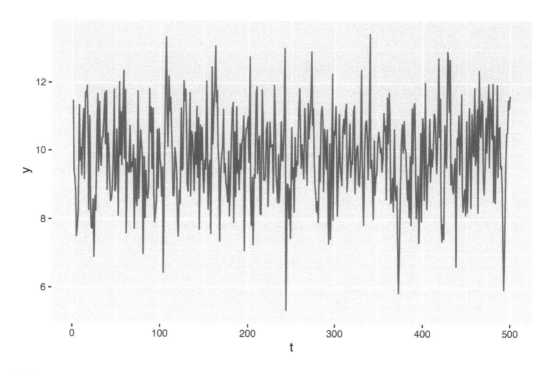

FIGURE 3.8: Simulated AR(2) with level plus noise series.

The 1-1 transformation in (3.9) can be written as

$$\phi_1 = r_1(1 - r_2), \text{ and } \phi_2 = r_2.$$

The model fitting is shown below.

```
y.ar2 <- sim.ar2$sim.data
n <- length(y.ar2)
id.x <- 1:n
ar2.dat <- cbind.data.frame(y.ar2, id.x)
formula.ar2 <- y.ar2 ~ f(id.x,
                    model = "ar",
                    order = 2,
                    constr = FALSE)
model.ar2 <- inla(
  formula.ar2,
  family = "gaussian",
  data = ar2.dat,
  control.predictor = list(compute = TRUE)
)
# summary(model.ar2)
format.inla.out(model.ar2$summary.hyperpar[,c(1:2)])
```

```
##    name                           mean   sd
## 1 Precision for Gaussian obs    0.808  0.062
## 2 Precision for id.x                   2.106  0.481
## 3 PACF1 for id.x                       0.856  0.021
## 4 PACF2 for id.x                      -0.743  0.074
format.inla.out(model.ar2$summary.hyperpar[,c(3:5)])
##    name                          0.025q  0.5q   0.975q
## 1 Precision for Gaussian obs    0.692   0.806   0.937
## 2 Precision for id.x             1.335   2.047   3.215
## 3 PACF1 for id.x                 0.812   0.857   0.893
## 4 PACF2 for id.x                -0.863  -0.752  -0.572
```

We can recover the summary of the fixed level α using

```
format.inla.out(model.ar2$summary.fixed[,c(1:5)])
##    name           mean   sd    0.025q  0.5q   0.975q
## 1 Intercept     9.843  0.067  9.712   9.843  9.974
```

The inverse of the 1-1 transformation (3.41) enables us to recover the estimated ϕ_1 and ϕ_2 coefficients from the estimated partial autocorrelations, using the following functions:

```
pacf <- model.ar2$summary.hyperpar$mean[3:4]
phi <- inla.ar.pacf2phi(pacf)
print(phi)
## [1]  1.491 -0.743
```

We see that these values $\hat{\phi}_1 = 1.491$ and $\hat{\phi}_2 = -0.743$ are close to the true values $\phi_1 = 1.5$ and $\phi_2 = -0.75$. We can recover the variance of the state error w_t as follows. Recall that the Wold representation (or the MA(∞) representation) of a stationary AR(p) process (3.6) is given by (see Shumway and Stoffer (2017))

$$x_t = \phi^{-1}(B)w_t = \sum_{j=0}^{\infty} \psi_j w_{t-j}, \qquad (3.10)$$

where the weights in the infinite order polynomial $\psi(B) = \sum_{j=0}^{\infty} \psi_j B^j$ are obtained by solving $\psi(B)\phi(B) = 1$. Then,

$$\text{Var}(x_t) = \sigma_w^2 \sum_{j=0}^{\infty} \psi_j^2. \qquad (3.11)$$

We use the R function ARMAtoMA() to get a large number (here, 100) of ψ weights to approximate the infinite sum in (3.11). The inla() output provides an estimate of $1/\text{Var}(x_t)$; using these, we can recover an estimate of σ_w^2 using (3.11).

```
phi <- c(1.5, -0.75)
psiwt.ar2 <- ARMAtoMA(ar = phi, ma = 0, 100)
precision.x.hat <- model.ar2$summary.hyperpar$mean[2]
sigma2.w.hat <-
  inla.emarginal(
    fun = function(x)
      exp(-x),
    marginal = model.ar2$internal.marginals.hyperpar$`Log precision for id`
```

```
  ) / sum(psiwt.ar2 ^ 2)
cat(paste(
  "Estimated state noise variance, sigma2.w",
  round(sigma2.w.hat, 2),
  sep = " = "
),
"\n")
## Estimated state noise variance, sigma2.w = 0.07
```

The posterior estimate 0.07 is close to the true value of $\sigma_w^2 = 0.05$.

We end this example by estimating posterior marginals of the parameters ϕ_1 and ϕ_2 using `inla.hyperpar.sample()`. In Figure 3.9, we show plots of the marginal posterior distributions of ϕ_1 and ϕ_2.

```
n.samples <- 10000
pacfs <- inla.hyperpar.sample(n.samples, model.ar2)[, 3:4]
phis <- apply(pacfs, 1L, inla.ar.pacf2phi)
par(mfrow = c(1, 2))
plot(
  density(phis[1, ]),
  type = "l",
  main="",
  xlab = expression(phi[1]),
  ylab = "density"
)
abline(v = 1.5)
plot(
  density(phis[2, ]),
  type = "l",
  main="",
  xlab = expression(phi[2]),
  ylab = "density"
)
abline(v = -0.75)
```

Since R-INLA parametrizes ϕ_1 and ϕ_2 as r_1 and r_2 (see (3.9)), we show the code to obtain their marginal posteriors as well (see Figure 3.10).

```
par(mfrow = c(1, 2))
plot(
  model.ar2$marginals.hyperpar$`PACF1 for id.x`,
  type = "l",
  main = "",
  xlab = expression(r[1]),
  ylab = "density"
)
plot(
  model.ar2$marginals.hyperpar$`PACF2 for id.x`,
  type = "l",
  main = "",
  xlab = expression(r[2]),
```

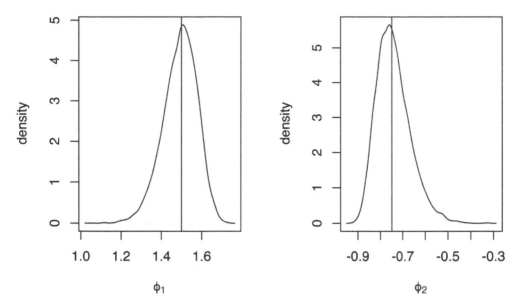

FIGURE 3.9: Marginal posterior densities of ϕ_1 (left panel) and ϕ_2 (right panel) in the AR(2) with level plus noise model. The vertical lines represent the true values assumed in the simulation.

```
   ylab = "density"
)
```

We end this section with a note that the syntax for fitting an AR(p) model also works when $p = 1$, so that we can also fit an AR(1) with the level plus noise model using

```
formula <- y ~ f(id.x, model = "ar",order = 1)
```

3.5 Random walk with drift plus noise model

The state equation describes x_t as a random walk with drift δ plus Gaussian noise. We then observe the signal x_t with measurement noise, as described below:

$$y_t = x_t + v_t; \quad v_t \sim N(0, \sigma_v^2), \tag{3.12}$$

$$x_t = \delta + x_{t-1} + w_t; \quad w_t \sim N(0, \sigma_w^2). \tag{3.13}$$

Using the `arima.sim()` function, we simulate the latent Gaussian process x_t, assuming a state error variance $\sigma_w^2 = 0.05$ and drift $\delta = 0.1$. Adding Gaussian observation noise v_t with variance $\sigma_v^2 = 0.5$ then generates y_t.

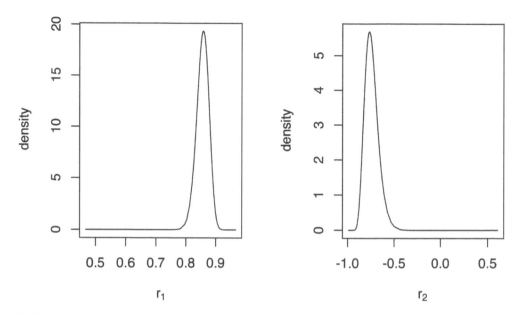

FIGURE 3.10: Marginal posterior densities of PACF(1) and PACF(2) in the internal R-INLA representation for an AR(2) with level plus noise model.

Since it is not possible to directly code the drift δ into the R-INLA formula, we must *reparameterize* the model as follows. Let z_t denote a random walk process without drift, derived from the same Gaussian white noise w_t. We can rewrite the random walk plus drift process x_t in (3.13) as

$$x_t = \delta t + z_t, \tag{3.14}$$
$$z_t - z_{t-1} = w_t, \tag{3.15}$$

so that

$$x_t - x_{t-1} = \delta t + z_t - \delta(t-1) - z_{t-1} = \delta + (z_t - z_{t-1}) = \delta + w_t.$$

Substituting x_t from (3.15) into the observation equation (3.12), we get the model

$$y_t = \delta t + z_t + v_t; \quad v_t \sim N(0, \sigma_v^2), \tag{3.16}$$
$$z_t = z_{t-1} + w_t; \quad w_t \sim N(0, \sigma_w^2). \tag{3.17}$$

That is, we have re-written a DLM with a drift coefficient in the state equation (see (3.13)) as a DLM with a deterministic trend δt in the observation equation (3.16), and a random walk without drift in the state equation (see (3.17)). This *reparametrization* of the DLM now enables us to formulate and fit the random walk with drift plus noise model in R-INLA. To illustrate, we generate a time series from this model; see Figure 3.11.

```
sim.rw1.drift <-
  simulation.from.rw1(
    sample.size = 500,
    burn.in = 100,
    level = 0,
    drift = 0.1,
    V = 0.5,
    W = 0.05,
    plot.data = TRUE,
    seed = 1
  )
sim.rw1.drift$sim.plot
```

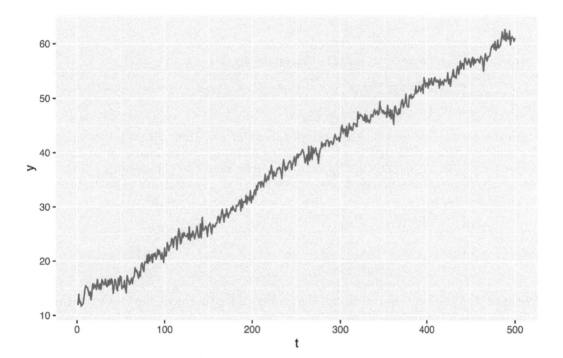

FIGURE 3.11: Simulated data from a random walk with drift plus noise model.

There are two ways to implement the modeling in R-INLA.

Method 1. Let *id.delta* correspond to the non time-varying coefficient δ in (3.16), taking values from 1 to n. Let *id.w* be an n-dimensional vector denoting the white noise w_t which corresponds to z_t (the random walk process without drift defined in (3.17)). We assume the prior distribution of δ to be $N(\text{mean} = 0, \text{precision} = 0.001)$, the default prior specification, which is a vague prior specification on δ. This can be accessed from `inla.set.control.fixed.default()[c("mean", "prec")]`. The priors for $\theta_v = \log(1/\sigma_v^2)$ and $\theta_w = \log(1/\sigma_w^2)$ have been described earlier. Based on the time indexes *id.w* and *id.delta*, we create a data frame *rw1.drift.dat* with three columns containing the time series y_t and the time indexes. The formula, model setup, and results are shown below.

```
y.rw1.drift <- sim.rw1.drift$sim.data
n <- length(y.rw1.drift)
id.w <- id.delta <- 1:n
rw1.drift.dat <- cbind.data.frame(y.rw1.drift, id.w, id.delta)

formula.rw1.drift.1 <-
  y.rw1.drift ~ f(id.w, model = "rw1", constr = FALSE) - 1 + id.delta
model.rw1.drift.1 <- inla(
  formula.rw1.drift.1,
  family = "gaussian",
  data = rw1.drift.dat,
  control.predictor = list(compute = TRUE)
)
# summary(model.rw1.drift.1)
format.inla.out(model.rw1.drift.1$summary.fixed[,c(1:5)])
##     name      mean  sd   0.025q  0.5q  0.975q
## 1 id.delta 0.098 0.01 0.079  0.098 0.117
format.inla.out(model.rw1.drift.1$summary.hyperpar[,c(1:2)])
##     name                        mean    sd
## 1 Precision for Gaussian obs  1.717 0.125
## 2 Precision for id.w          23.676 5.893
```

From the output, we see that the posterior mean of δ is 0.098 with a posterior standard deviation of 0.01; recall that the true value of δ is 0.1. The estimates of σ_v^2 and σ_w^2 are respectively 0.582 (true value is 0.5) and 0.042 (true value is 0.05). Note that if we wish to change the prior specification on δ, we can use `control.fixed` within `inla()` (although we do not show this here).

```
prior.fixed <- list(mean = 0, prec = 0.1)
inla(formula = ,
     data = ,
     control.fixed = prior.fixed)
```

Method 2. This is an equivalent R-INLA approach for this model and uses a formulation with the `model = linear` option.

```
formula.rw1.drift.2 <-
  y.rw1.drift ~ f(id.w, model = "rw1", constr = FALSE) - 1 +
  f(id.delta, model = "linear", mean.linear = 0, prec.linear = 0.001)
model.rw1.drift.2 <- inla(
  formula.rw1.drift.2,
  family = "gaussian",
  data = rw1.drift.dat,
  control.predictor = list(compute = TRUE)
)
# summary(model.rw1.drift.2)
format.inla.out(model.rw1.drift.2$summary.fixed[,c(1:5)])
##     name      mean  sd   0.025q  0.5q  0.975q
## 1 id.delta 0.098 0.01 0.079  0.098 0.117
format.inla.out(model.rw1.drift.2$summary.hyperpar[,c(1:2)])
##     name                        mean    sd
```

```
## 1 Precision for Gaussian obs   1.717 0.125
## 2 Precision for id.w              23.676 5.893
```

Note that we get similar results from both methods, and we may use either code to fit and analyze this model. Similar to what we have seen earlier, the following code is useful to recover and print posterior estimates of σ_v^2, σ_w^2 and δ from the `inla()` output.

```
sigma2.v.hat <-
  inla.emarginal(
    fun = function(x)
      exp(-x),
    marginal = model.rw1.drift.1$internal.marginals.hyperpar$
      `Log precision for the Gaussian observations`
  )
sigma2.w.hat <-
  inla.emarginal(
    fun = function(x)
      exp(-x),
    marginal = model.rw1.drift.1$internal.marginals.hyperpar$
      `Log precision for id.w`
  )
cat(paste(
  "Estimated observation noise variance, sigma2.v",
  round(sigma2.v.hat, 2),
  sep = " = "
),
"\n")
## Estimated observation noise variance, sigma2.v = 0.59
cat(paste(
  "Estimated state noise variance, sigma2.w",
  round(sigma2.w.hat, 2),
  sep = " = "
),
"\n")
## Estimated state noise variance, sigma2.w = 0.04
delta.hat <- model.rw1.drift.1$summary.fixed$mean
cat(paste("Estimated drift, delta", round(delta.hat, 2), sep = " = "), "\n")
## Estimated drift, delta = 0.1
```

Also, as we discussed earlier, we can pull out the posterior distributions of the hyperparameters using `inla.tmarginal()` and obtain HPD intervals using `inla.hpdmarginal()` (code not shown here). We will revisit Method 2 in Chapter 5, further explaining the available options related to the term *linear* in the formula. We next describe an approach to use R–INLA for modeling a DLM with a *time-varying drift* in the state equation.

3.6 Second-order polynomial model

Given an observed time series y_t, $t = 1, \ldots, n$, consider the model with a random walk latent state x_t with a random, time-varying drift δ_{t-1}:

$$y_t = x_t + v_t, \tag{3.18}$$
$$x_t = x_{t-1} + \delta_{t-1} + w_{t,1}, \tag{3.19}$$
$$\delta_t = \delta_{t-1} + w_{t,2}. \tag{3.20}$$

This is usually called the second-order polynomial model. Similar models have been discussed in Ruiz-Cárdenas et al. (2012).

Let $\boldsymbol{x}_t^* = (x_t, \delta_t)'$ be the 2-dimensional state vector with $x_{t,1}^* = x_t$ and $x_{t,2}^* = \delta_t$. For $t = 2, \ldots, n$, we can write (3.19) and (3.20) as

$$\boldsymbol{x}_t^* = \begin{pmatrix} 1 & 1 \\ 0 & 1 \end{pmatrix} \boldsymbol{x}_{t-1}^* + \boldsymbol{w}_t^*$$
$$= \boldsymbol{\Phi}\boldsymbol{x}_{t-1}^* + \boldsymbol{w}_t^*, \text{ say}, \tag{3.21}$$

where $\boldsymbol{\Phi} = \begin{pmatrix} 1 & 1 \\ 0 & 1 \end{pmatrix}$ is a 2×2 state transition matrix. We assume that $\boldsymbol{w}_t^* \sim N(\boldsymbol{0}, \boldsymbol{W})$, where $\boldsymbol{W} = \mathrm{diag}(\sigma_{w1}^2, \sigma_{w2}^2)$. The observation equation in the model is a linear function of the bivariate state vector $\boldsymbol{x}_t^*$ plus an observation noise v_t:

$$y_t = (1, \ 0)\boldsymbol{x}_t^* + v_t, \ t = 1, \ldots, n, \tag{3.22}$$

where $v_t \sim N(0, V)$.

Following Ruiz-Cárdenas et al. (2012), we can express (3.21) as

$$\boldsymbol{0} = \boldsymbol{x}_t^* - \boldsymbol{\Phi}\boldsymbol{x}_{t-1}^* - \boldsymbol{w}_t^*, \ t = 2, \ldots, n, \tag{3.23}$$

yielding a set of $n - 1$ *faked zero observations* on the left side written as a linear function of the terms on the right side. That is, we rewrite the state equations (3.19) and (3.20) as

$$0 = x_t - x_{t-1} - \delta_{t-1} - w_{t,1}, \tag{3.24}$$
$$0 = \delta_t - \delta_{t-1} - w_{t,2}. \tag{3.25}$$

Then, the augmented model has dimension $n + 2(n - 1)$, the n observations $y_1, \ldots, y_n$ being stacked together with the set of $n - 1$ faked zero observations in (3.23):

$$\begin{pmatrix} y_1 & NA & NA \\ \vdots & \vdots & \vdots \\ y_n & NA & NA \\ NA & 0 & NA \\ \vdots & \vdots & \vdots \\ NA & 0 & NA \\ NA & NA & 0 \\ \vdots & \vdots & \vdots \\ NA & NA & 0 \end{pmatrix}, \tag{3.26}$$

where each of the second and third horizontal blocks has dimension $(n - 1) \times 3$. The first n rows of Column 1 contain the observations, $y_1, \ldots, y_n$. Column 2 is associated with the first state variable $x_{t,1}^* = x_t$ whose entries are forced to zero for $t = 2, \ldots, n$ (see (3.23)). Column 3 is associated with the second state variable $x_{t,2}^* = \delta_t$ for $t = 2, \ldots, n$. All the other entries in the structure have no values and are assigned NAs. In summary, this augmented structure in (3.26) will have the observed y_t in the first n rows of Column 1, and there will be as many additional columns as the number of state variables in the state equation (here, 2).

For fitting the model, R-INLA considers different likelihoods for Column 1 and the other columns of the augmented structure in (3.26). As described in Ruiz-Cárdenas et al. (2012), given the state vector x_t^*, the data y_t, $t = 1, \ldots, n$ are assumed to follow a Gaussian distribution with unknown precision V^{-1}. Given x_t^*, x_{t-1}^* and w_t^*, the faked zero data in Columns 2 and 3 of (3.26) are artificial observations and are assumed to be deterministic (i.e., known, with zero variance). R-INLA represents this condition by assuming that these faked zero data follow a Gaussian distribution with a high, fixed precision. It is assumed that for $t = 1$, x_1^* follows a noninformative distribution, i.e., a bivariate normal distribution with a fixed and low-valued precision matrix. For $t = 2, \ldots, n$, it is assumed that x_t^* follows (3.21) and $w_t^* \sim N(0, W)$:

$$f_{w_t^*}(x_t^* - \Phi x_{t-1}^*) \propto \exp[-\frac{1}{2}(x_t^* - \Phi x_{t-1}^*)' W^{-1}(x_t^* - \Phi x_{t-1}^*)].$$

We must define the $n \times 2$ matrix $X^* = (x_1^*, \ldots, x_n^*)'$, and also define $X_{(-1)}^* = (x_2^*, \ldots, x_n^*)'$ and $X_{(-n)}^* = (x_1^*, \ldots, x_{n-1}^*)'$.

The following code simulates a time series y_t, see Figure 3.12, of length $n = 200$. The true values of the error variances are assumed to be $V = 0.01$, and $W = \text{diag}(0.0001, 0.0001)$.

```
sim.secondorder <-
  simulation.from.second.order.dlm(
    sample.size = 200,
    V = 0.01,
    W1 = 1e-04,
    W2 = 1e-04,
    plot.data = TRUE,
    seed = 123457
  )
sim.secondorder$sim.plot

y.sec <- sim.secondorder$sim.data
n <- length(y.sec)
```

The following code sets up the augmented model. Let $m = n - 1$. The observed time series y_t forms the first block of rows.

```
m <- n-1
Y <- matrix(NA, n+2*m, 3)
# actual observations (y.sec)
Y[1:n, 1] <- y.sec
Y[1:m + n, 2] <- 0
Y[1:m + (n+m), 3] <- 0
```

The set up of the vectors of indexes and vectors of weights (coefficients) for the different latent variables follows the discussion in Ruiz-Cárdenas et al. (2012) (also see their Appendix). In order to avoid confusion with the state errors w_t, we use the symbols ci, cj, etc., instead of wi, wj, etc., to denote the weights (coefficients). We give a description of these items for this model.

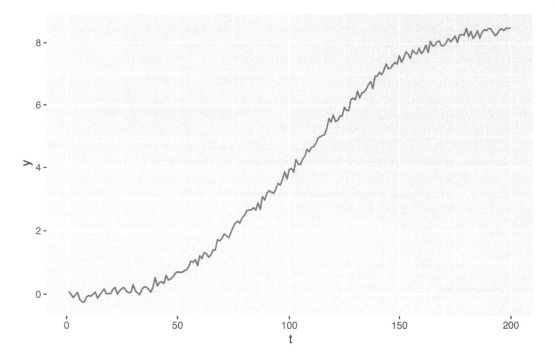

FIGURE 3.12: Simulated data from a second-order polynomial model.

1. The vector of indexes denoted by $\boldsymbol{i}$ for the latent series $x^*_{t,1} = x_t$ consists of (a) a vector $c(1:n)$ of length n (from the observation equation), (b) a vector $c(2:n)$ of length m (from the rewritten first state equation (3.24) for $x^*_{t,1} = x_t$ with fake zeroes), and (c) a vector of NA's of length m (from the rewritten second state equation (3.25) for $x^*_{t,2} = \delta_t$ with fake zeroes).

2. The vector of indexes $\boldsymbol{j}$ for the latent series $x^*_{t-1,1} = x_{t-1}$ consists of (a) a vector of NA's of length n (since the observation equation has no x_{t-1}), (b) a vector $c(1:m)$ of length m (from the rewritten state equation (3.24)), and (c) a vector of NA's of length m (since the rewritten second state equation (3.25) has no x_{t-1}).

3. The vector $\boldsymbol{cj}$ consists of the coefficients corresponding to the vector $\boldsymbol{j}$ and has (a) a vector of NA's of length n (since the observation equation has no x_{t-1}), (b) a vector of length m with elements -1 (corresponding to the coefficient of x_{t-1} in (3.24)), and (c) a vector of NA's of length m (since the second state equation has no x_{t-1}).

4. The vector of indexes $\boldsymbol{l}$ for the latent series $x^*_{t,2} = \delta_t$ consists of (a) a vector of NA's of length n (since the observation equation has no δ_t), (b) a vector of NA's of length m (since the first state equation has no δ_t), and (c) a vector $c(2:n)$ of length m (corresponding to the coefficient of δ_t in the rewritten second state equation).

5. The vector $\boldsymbol{cl}$ consists of coefficients corresponding to the vector $\boldsymbol{l}$ and has (a) a vector of NA's of length n (since the observation equation has no δ_t), (b) a vector of NA's of length $m - 1$ (since the first rewritten state equation (3.24) has no δ_t),

and (c) a vector of length m with elements 1 (corresponding to the coefficient of δ_t in (3.25)).

6. The vector of indexes $\boldsymbol{k}$ for the latent series $x^*_{t-1,2} = \delta_{t-1}$ consists of (a) a vector of NA's of length n (since the observation equation has no δ_{t-1}), (b) a vector $c(1:m)$ of length m (from the rewritten state equation (3.24)), and (c) a vector $c(1:m)$ of length m (from the rewritten state equation (3.24)).

7. The vector $\boldsymbol{ck}$ consists of coefficients corresponding to the vector of indexes $\boldsymbol{k}$ and has (a) a vector of NA's of length n (since the observation equation has no δ_{t-1}), (b) a vector of length m with elements -1 (corresponding to the coefficient of δ_{t-1} in (3.24)), and (c) a vector of length m with elements -1 (corresponding to the coefficient of δ_{t-1} in (3.25)).

8. The vector $\boldsymbol{q}$ consists of indexes for the state errors (perturbations) $w_{t,1}$ in (3.24) and has NA's in all entries except for the m middle entries corresponding to (3.24).

9. The vector $\boldsymbol{cq}$ correspondingly has -1 in the middle m positions and NA's otherwise.

10. The vector $\boldsymbol{s}$ consists of indexes for the state errors (perturbations) $w_{t,2}$ in (3.25) and has NA's in all entries except for the last m entries corresponding to (3.25).

11. The vector $\boldsymbol{cs}$ correspondingly has -1 in the last m positions and NA's otherwise.

```
i <- c(1:n, 2:n, rep(NA, m))           # indexes for x_t
j <- c(rep(NA, n), 2:n - 1, rep(NA, m))    # indexes for x_{t-1}
cj <- c(rep(NA, n), rep(-1, m), rep(NA, m)) # weights for x_{t-1}
l <- c(rep(NA, n + n - 1), 2:n)            # indexes for delta_t
cl <- c(rep(NA, n + m), rep(1, m))         # weights for delta_t
k <- c(rep(NA, n), 2:n - 1, 2:n - 1)   # indexes for delta_{t-1}
ck <- c(rep(NA, n), rep(-1, 2 * m))        # weights for delta_{t-1}
q <- c(rep(NA, n), 1:m, rep(NA, m))    # indexes for w_{t,1}
s <- c(rep(NA, n + m), 1:m)            # indexes for  w_{t,2}
cq <- c(rep(NA, n), rep(-1, m), rep(NA, m)) # weights for w_{t,1}
cs <- c(rep(NA, n + m), rep(-1, m))        # weights for w_{t,2}
```

The model formulation is shown below. The first term handles the first state variable x_t (with index vector $\boldsymbol{i}$) while the second term handles the state variable x_{t-1} (with index vector $\boldsymbol{j}$). Since both have the same distribution, we may use `f(j, wj, copy="i")`. A similar usage follows for index vectors $\boldsymbol{l}$ and $\boldsymbol{k}$ for the second state variables δ_t and δ_{t-1}. The R-INLA feature `copy` enables us to share terms among different parts of the model and is explained below (also see Section 6.5 of Gómez-Rubio (2020)).

R-INLA feature copy

Using the `copy` feature when defining different likelihoods associated with different latent random variables allows us to ensure that linear predictors of the different likelihoods share the same type of latent effect, with the values of the random effects being the same, while copied effects can be scaled by a parameter. All copies of the random effects share the same hyperparameters.

```
formula.second.poly = Y ~ f(i, model="iid", initial=-10, fixed=TRUE) +
            f(j, cj, copy="i") +
            f(l, cl, model="iid", values = 1:n, initial=-10, fixed=TRUE)+
```

```
f(k, ck, copy="1")+
f(q, cq, model ="iid") +
f(s, cs, model ="iid") -1
```

Here, i is the index variable used in the original effect, i.e., x_t, and j is the index corresponding to x_{t-1} which must take values in the same set as i. In general, the new index j need not have the same ordering as i, and only the elements of x_t that are indexed in j will be copied (as pointed out in Gómez-Rubio (2020)).

The argument `values` in the function `f()`is used to provide the set of all values assumed by the covariate for which we want the estimated effect. By default, `values` assumes the unique set of values defined in the index; for example, for index l, the set is $\{2,\ldots,n\}$. In the model formula for the index k, we have $copy =$ "1"; the argument `copy` enables the index k to inherit values of the argument `values` defined for index l. However, note that `values` for $f(k,ck,copy =$ "1") is the set $\{1,\ldots,n-1\}$. Therefore, for `inla()` to run without error, it is necessary to inform the model that l is being *copied*, for which we must include `values = 1:n` in the function `f()` defined for l.

```
model.second.poly = inla(
  formula.second.poly,
  data = data.frame(i, j, cj, k, ck, l, q, s, cq, cs),
  family = rep("gaussian", 3),
  control.family = list(
    list(),
    list(initial = 10, fixed = TRUE),
    list(initial = 10, fixed = TRUE)
  ),
  control.predictor = list(compute = TRUE)
)
format.inla.out(model.second.poly$summary.hyperpar[,c(1:2)])
##   name                         mean   sd
## 1 Precision for Gaussian obs    105   11.25
## 2 Precision for q             20081 16381.84
## 3 Precision for s             51361 26504.35

post.mean.V <- inla.emarginal(
  fun = function(x)
    exp(-x),
  marginal = model.second.poly$internal.marginals.hyperpar$
    `Log precision for the Gaussian observations`
)
post.mean.W11 <- inla.emarginal(
  fun = function(x)
    exp(-x),
  marginal = model.second.poly$internal.marginals.hyperpar$
    `Log precision for s`
)
post.mean.W22 <- inla.emarginal(
  fun = function(x)
    exp(-x),
  marginal = model.second.poly$internal.marginals.hyperpar$
```

```
    `Log precision for q`
)
```

The posterior mean of V is 0.0096. The posterior means of σ_{w1}^2 and σ_{w2}^2 are 0.000025 and 0.00009833 respectively, and are not recovered as accurately as the mean of V.

3.7 Forecasting states and observations

An important goal in time series analysis is to forecast future values of a time series based on a fitted model. Suppose we split the observed data into a fitting or training portion $y_1, \ldots, y_{n_t}$ and a holdout or test portion $y_{n_t+1}, \ldots, y_{n_t+n_h}$. The training portion is of length n_t, and the holdout portion is of length n_h. In the Bayesian DLM framework, given the training data $y_1, \ldots, y_{n_t}$, estimates of the states $x_1, \ldots, x_{n_t}$ and estimates of the model hyperparameters $\boldsymbol{\theta}$, we first forecast $x_{n_t+\ell}$, $\ell = 1, \ldots, n_h$ by $f(x_{n_t+\ell}|\boldsymbol{y}^n)$, $\ell = 1, \ldots, n_h$. We then obtain n_h forecasts of $y_{n_t+\ell}$ by $f(y_{n_t+\ell}|\boldsymbol{y}^n)$, $\ell = 1, \ldots, n_h$.

In R-INLA, there is no separate forecasting function like the **predict()** function in the **stats** package which is commonly used with functions such as **lm()**. Therefore, we append the end of the training data of length n_t with NA's to represent the times to hold out the last n_h data points. Forecasts for $x_{n_t+\ell}$ are obtained using the **summary.random** option. For example, we can forecast $n_h = 10$ future states in the AR(1) with level plus noise model; we use the **summary.random** option, as shown below.

```
format.inla.out(tail(model.ar1.level$summary.random$id.x[,c(1:3)],10))
##      ID   mean    sd
## 1   491   0.105  0.272
## 2   492  -0.169  0.273
## 3   493  -0.437  0.300
## 4   494  -0.135  0.273
## 5   495   0.220  0.279
## 6   496   0.239  0.275
## 7   497   0.122  0.272
## 8   498   0.295  0.276
## 9   499   0.261  0.275
## 10  500   0.259  0.281
```

We then obtain $n_h = 10$ future forecasts $y_{n_t+\ell}$ using

```
format.inla.out(tail(model.ar1.level$summary.fitted.values[,c(1:3)],10))
##      name           mean   sd     0.025q
## 1   fit.Pred.491   1.392  0.271  0.864
## 2   fit.Pred.492   1.118  0.273  0.575
## 3   fit.Pred.493   0.850  0.300  0.243
## 4   fit.Pred.494   1.152  0.272  0.611
## 5   fit.Pred.495   1.507  0.278  0.972
## 6   fit.Pred.496   1.526  0.275  0.996
## 7   fit.Pred.497   1.409  0.272  0.872
## 8   fit.Pred.498   1.583  0.276  1.050
```

```
## 9  fit.Pred.499 1.549 0.274 1.016
## 10 fit.Pred.500 1.547 0.281 1.002
```

Figure 3.13 shows the time series generated by the AR(1) with level plus noise model, together with the last 10 forecasts. For a better visualization of forecasts (means of the posterior predictive distributions) and 2 standard deviation bounds, we show a plot with only the last 100 data points.

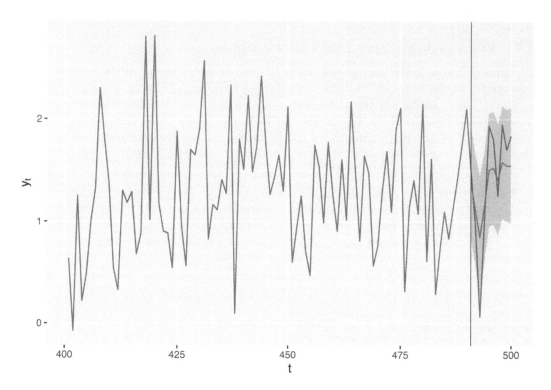

FIGURE 3.13: Forecasts (blue) for ten future time points along with observed data (red) from the AR(1) with level plus noise model. The black vertical line divides the data into training and test portions. The gray shaded area shows forecast intervals based on 2 standard deviation (sd) bounds around the means of the posterior predictive distributions.

3.8 Model comparisons

In this section, we explain how to compare different fitted models. Many model comparison criteria have been defined in the Bayesian framework, including the conditional predictive ordinate (CPO), the Bayes factor (BF), the deviance information criterion (DIC), and the Watanabe Akaike information criterion (WAIC). These are useful for comparing the relative performance of the fitted models based on the calibration (or training or fitting) data, and were reviewed in Chapters 1 and 2. We may also compare models based on their predictive performance using frequentist metrics such as the mean absolute percent error (MAPE) or mean absolute error (MAE). These two metrics can be computed from in-sample training data as well as out-of-sample test data.

3.8.1   In-sample model comparisons

We start with a discussion of in-sample model comparisons and illustrate code for the random walk plus noise model. In R-INLA, we compute these quantities using the `control.compute` option in the formula. If the output is saved in `out`, we can access the predictive quantities as `out$dic$dic`, or `out$cpo$cpo`, or `out$cpo$pit`, as shown below.

1. The *deviance information criterion* (DIC) was defined in (1.10). We use the `control.compute` option to obtain the DIC, via `control.compute=list(dic=TRUE)`, and access it by `out$dic$dic`.

```
model.rw1 <- inla(
  formula.rw1,
  family = "gaussian",
  data = rw1.dat,
  control.predictor = list(compute = TRUE),
  control.compute = list(dic = TRUE, cpo = TRUE)
)
model.rw1$dic$dic
## [1] 1312
```

2. The *conditional predictive ordinate* (CPO) was defined in Chapter 1 as $p(y_t|\boldsymbol{y}_{(-t)})$, a "leave-one-out" measure of model fit. The sum of the log CPO values gives us the pseudo-Bayes factor. We again use `control.compute=list(cpo=TRUE)`, and access it by `out$cpo$cpo`.

```
cpo.rw1 <- model.rw1$cpo$cpo
psBF.rw1 <- sum(log(cpo.rw1))
psBF.rw1
## [1] -671
```

3. The *probability integral transform* (PIT) defined by

$$p(y_t^{\text{new}} \leq y_t|\boldsymbol{y}_{(-t)}) \tag{3.27}$$

is another "leave-one-out" measure of fit computed by R-INLA. A histogram of PIT must resemble a uniform distribution; extreme values indicate outlying observations. We again use `control.compute=list(cpo=TRUE)`, and access by using `out$cpo$pit`. The plot in Figure 3.14 gives an indication of model adequacy, since the histogram is close to that of a sample from a uniform distribution.

```
pit.rw1 <- model.rw1$cpo$pit
hist(pit.rw1,main =" ", ylab="frequency")
```

Note that R-INLA computes CPO or PIT by keeping the integration-points fixed. To get improved grid integration, we can include, for instance, `control.inla=list(int.strategy = "grid", diff.logdens = 4)`. Also, to increase the accuracy of the tails of the marginals, we can use `control.inla = list(strategy = "laplace", npoints = 21)`. This adds 21 evaluation points instead of the default `npoints=9`.

It is useful to understand what the following check for failure means. R-INLA makes some implicit assumptions which are tested using internal checks. If `model.rw1$cpo$failure[i]`

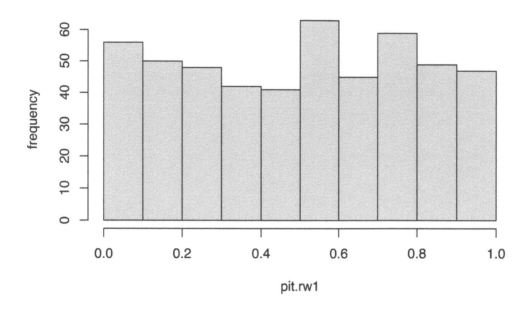

FIGURE 3.14: Histogram of the probability integral transform (PIT) for the random walk plus noise model.

== 0, then no assumption is violated. If `model.rw1$cpo$failure[i] > 0` for any row i of the data, this indicates that some assumption is violated, larger values (maximum value is 1) indicating more serious violations of assumptions.

```
table(model.rw1$cpo$failure)
##
##    0
## 500
```

It might be a good idea to recompute those, if any, CPO/PIT values for which `result$cpo$failure>0`. INLA has provided `improved.result = inla.cpo(result)`. This takes an object which is the output from `inla()`, and recomputes (in an efficient way) the CPO/PIT for which `result$cpo$failure > 0`, and returns `model.rw1` with the improved estimates of CPO/PIT. See `?inla.cpo` for details. R-INLA recommends caution when using CPO or PIT for model validation, since they may depend on tail behavior, which is difficult to estimate. We can verify that the smallest CPO values are estimated correctly by setting `res$cpo$failure[smallest cpo indices] = 1` and re-estimating them with `inla.cpo()`.

4. R-INLA also outputs the marginal likelihood (see the definition in Chapter 1) on the logarithmic scale by default. Recall that a model with higher marginal likelihood is preferred.

```
model.rw1$mlik
##                                                    [,1]
```

```
## log marginal-likelihood (integration) -750.0
## log marginal-likelihood (Gaussian)    -750.3
```

3.8.2   Out-of-sample comparisons

Let $y_{n_t}(\ell)$ denote the forecast from origin n_t for $y_{t+\ell}$, $\ell = 1, \ldots, n_h$, and let $e_{n_t}(\ell) = y_{n_t+\ell} - y_{n_t}(\ell)$ be the ℓth forecast error. The mean absolute percent error (MAPE) is a well known criterion, defined by

$$\text{MAPE} = \frac{100}{n_h} \sum_{\ell=1}^{n_h} |e_{n_t}(\ell)/y_{n_t+\ell}|. \tag{3.28}$$

Another widely used criterion is the mean absolute error (MAE) given by

$$\text{MAE} = \frac{1}{n_h} \sum_{\ell=1}^{n_h} |e_{n_t}(\ell)|. \tag{3.29}$$

Small values of MAPE and MAE indicate better fitting models. We show the custom code for computing MAPE and MAE in Chapter 14. We can also compute in-sample MAPE and in-sample MAE, as we discuss in the following chapters.

3.9   Non-default prior specifications

We look at a few ways for specifying non-default priors for hyperparameters. A listing of all prior distributions that R-INLA supports is given in Table 14.5; this is similar to Table 5.2 in Gómez-Rubio (2020). In some situations, we may wish to use alternatives to the default prior specifications, selecting one of the other options available in R-INLA. We show this using $\theta_1 = \log(1/\sigma_w^2)$ and $\theta_2 = \log(\frac{1+\phi}{1-\phi})$.

1. We can change the hyperprior *parameter values* from their default settings. For example, for the log-gamma prior for $\log(1/\sigma_w^2)$, we can change `param = c(1, 0.00005)` to `param = c(1, 0.0001)` and/or `initial = 4` to `initial = 2`. The prior for θ_2 will remain at its default setting.

```
hyper.state = list(
  theta1 = list(
    prior = "loggamma",
    param = c(1, 0.00005),
    initial = 4,
    fixed = FALSE
  )
)
```

2. Or, we may use other prior distributions, such as the *beta correlation prior* for θ_2 (see `inla.doc("betacorrelation")` for more details on this prior). The prior for θ_1 will remain at its default setting.

```
hyper.state = list(
  theta2 = list(
    prior = "betacorrelation",
    param = c(2, 2),
    initial = -1.098,
    fixed = FALSE
  )
)
```

The betacorrelation prior is a prior for the AR(1) correlation parameter ϕ (recall this is denoted by ρ in R-INLA), represented internally as

$$\theta_{\phi,bc} = \log \frac{1+\phi}{1-\phi}. \tag{3.30}$$

This prior is defined on $\theta_{\phi,bc}$ so that ϕ is scaled to be in $(-1,1)$ and follow a Beta(a,b) distribution given by

$$\pi(\phi; a, b) = \frac{1}{2} \frac{\Gamma(a+b)}{\Gamma(a)\Gamma(b)} \phi^{a-1} (1-\phi)^{b-1}.$$

The inverse transformation from $\theta_{\phi,bc}$ to ϕ is the same as for the default prior on ϕ, and is done by R-INLA, so that the output is provided on ϕ.

3.9.1   Custom prior specifications

In some examples, we may wish to specify priors that are not included in the list of priors provided in R-INLA and listed in names(inla.models()$prior). In such cases, there are two ways for building user-defined priors, i.e., *table priors*, which are based on tabulated values from the desired densities and *expression priors*, which are specified by formulas for the desired prior density functions.

Table Priors

The user provides a table with values of the hyperparameter defined on R-INLA's internal scale and the associated prior densities on the log-scale. Values not defined in the table are interpolated at run time. INLA fits a spline through the points and continues with this in the succeeding computations. There is no transformation into a functional form. The input-format for the table is a string, which starts with table: and is then followed by a block of x-values and a block of corresponding y-values, which represent the values of the log-density evaluated at x. The syntax is:

```
table: x_1 ... x_n y_1 ... y_n
```

For more details, see inla.doc("table"). As an example, suppose we wish to set up a Gaussian prior with zero mean and precision 0.001 for a hyperparameter θ (on the internal scale). The following code sets this up as a table prior.

```
theta <- seq(-100, 100, by = 0.1)
log_dens <- dnorm(theta, 0, sqrt(1 / 0.001), log = TRUE)
gauss.prior <- paste0("table: ",
                 paste(c(theta, log_dens), collapse = " ")
)
```

Depending on what θ represents, `gauss.prior` will be included either in `formula` under `f()` or in `control.family`, as mentioned earlier.

Expression Priors

The expression prior definition uses the `muparser` library.[1] To set up a $N(0, 0.01)$ prior on a hyperparameter θ (i.e., with precision 100), we can include the following code in the formula and model execution:

```
gauss.pr.e <- "expression:
  mean = 0;
  prec = 1000;
  logdens = 0.5 * log(prec) - 0.5 * log (2 * pi);
  logdens = logdens - 0.5 * prec * (theta - mean)^2;
  return(logdens);
"
```

`R-INLA` recognizes several well known functions that can be used within the expression statement. These include common mathematical functions, such as $\exp(x)$, $\sin(x)$, x^y, $\mathrm{pow}(x; y)$ (to denote x raised to power y), $\mathrm{gamma}(x)$ and $\mathrm{lgamma}(x)$ (to denote the gamma function and its logarithm), etc.

3.9.2 Penalized complexity (PC) priors

`R-INLA` provides a useful framework for building penalized complexity (PC) priors (Fuglstad et al., 2019). These priors are defined on individual model components that can be regarded as flexible extensions of a simple and interpretable base model, and then penalizes deviations from the base model. This allows us to control flexibility, reduce over-fitting, and improve predictive performance. PC priors include a single parameter to control the degree of flexibility allowed in the model. They are specified by setting values (U, α) such that $P(T(\xi) > U) = \alpha$, where $T(\xi)$ is an interpretable transformation of the flexibility parameter xi, and U is an upper bound that specifies a tail event with probability α.

For example, recall that we assumed an inverse log-Gamma prior on σ_v^2. Instead, we can set a PC prior on the standard deviation σ_v by $P(\sigma_v > 1) = 0.01$:

```
prior.prec <- list(prec = list(prior = "pc.prec",
                               param = c(1, 0.01)))
```

Documentation and details on the PC prior `pc.prec` can be seen in `inla.doc("pc.prec")`.

3.10 Posterior sampling of latent effects and hyperparameters

In addition to producing posterior summaries as we have seen earlier, `R-INLA` can also generate samples from the approximate posterior distributions of the latent effects and the hyperparameters (Rue and Held, 2005; Gómez-Rubio and Rue, 2018). To do this, we first fit the `inla` model with the option `config = TRUE` in `control.compute`, which keeps the internal GMRF representation for the latent effects in the object returned by `inla()`. The table

---

[1] https://GitHub.com/beltoforion/muparser

below shows the options a user may specify with the function `inla.posterior.sample()` (see Table 2.11 in Gómez-Rubio (2020)).

We illustrate posterior sample generation for the AR(1) with level plus noise model, which we discussed earlier in this chapter. Let n denote the length of the entire time series, while n_t and n_h are the lengths of the training and test (or holdout) portions respectively. We use *n.train* and *n.hold* to denote these in the code. Note that we use test and holdout interchangeably in the book.

```
sim.ar1.level <-
  simulation.from.ar(
    sample.size = 500,
    burn.in = 100,
    phi = 0.6,
    level = 1.4,
    drift = 0,
    V = 0.2,
    W = 0.1,
    plot.data = FALSE,
    seed = 123457
  )
y.ar1.level <- sim.ar1.level$sim.data
n <- length(y.ar1.level)
n.hold <- 5
hold.y <- tail(y.ar1.level, n.hold)
train.y <- y.ar1.level[1:(length(y.ar1.level) - n.hold)]
# Append training data with NA
y <- train.y
y.append <- c(y, rep(NA, n.hold))
id.x <- 1:n
data.ar1.level <- cbind.data.frame(y.ar1.level, id.x)
formula.ar1.level <-
  y.ar1.level ~ f(id.x, model = "ar1", constr = FALSE)
model.ar1.level <- inla(
  formula.ar1.level,
  family = "gaussian",
  data = data.ar1.level,
  control.compute = list(config = TRUE),
  control.predictor = list(compute = TRUE)
)
# summary(model.ar1.level)
format.inla.out(model.ar1.level$summary.fixed[,c(1:5)])
##   name        mean   sd     0.025q 0.5q   0.975q
## 1 Intercept   1.287  0.034  1.221  1.287  1.354
format.inla.out(model.ar1.level$summary.hyperpar[,c(1:2)])
##   name                        mean   sd
## 1 Precision for Gaussian obs  4.509  0.591
## 2 Precision for id.x          8.170  1.853
## 3 Rho for id.x                0.450  0.075
```

Posterior samples for hyperparameters

We generate `n.samp = 2000` samples (default is 1000) from the posterior distributions of $1/\sigma_v^2, 1/\sigma_x^2$, and ϕ. Since the default option is `intern = FALSE`, these values are not in the internal R-INLA representations of these hyperparameters, but rather in their original representations. Recall that the true precisions are 5 and 6.4, while the true ϕ is 0.6. The `inla.hyperpar.sample` function employs the *grid integration strategy*, where the joint posterior of the hyperparameters is evaluated in a regular grid of points (see Rue et al. (2009) and Seppä et al. (2019)). The histograms of these posterior samples are symmetric, as seen in Figure 3.15.

```
n.samp <- 2000
set.seed(123457)
ar1.level.hyperpar.samples <-
  inla.hyperpar.sample(n.samp, model.ar1.level, improve.marginals = TRUE)
par(mfrow = c(2, 2))
hist(
  ar1.level.hyperpar.samples[, 2],
  xlab = expression(1/Var(x[t])),
  ylab = "frequency",
  main = "",
  sub = "(a)"
)
hist(
  ar1.level.hyperpar.samples[, 3],
  xlab = expression(phi),
  ylab = "frequency",
  main = "",
  sub = "(b)"
)
hist(
  ar1.level.hyperpar.samples[, 1],
  xlab = expression(1/sigma[v]^2),
  ylab = "frequency",
  main = "",
  sub = "(c)"
)
```

Next, we generate posterior samples for the latent random effects $x_t, t = 1, \ldots, n$. We set `num.threads=1`, and a random seed to facilitate reproducibility (see below for a discussion).

```
ar1.level.latent_hyper.samples <-
  inla.posterior.sample(n.samp,
                        model.ar1.level,
                        seed = 123457,
                        num.threads = 1)
length(ar1.level.latent_hyper.samples)
## [1] 2000
model.ar1.level$misc$configs$contents
## $tag
## [1] "Predictor"    "id.x"          "(Intercept)"
##
## $start
## [1]    1  501 1001
##
```

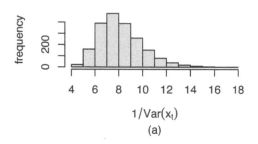

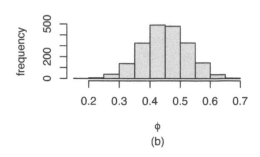

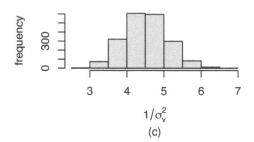

FIGURE 3.15: Histograms of samples of hyperparameters generated using the function inla.hyperpar.samples in the AR(1) with level plus noise model. Precision of state variable in (a), ϕ in (b), and precision of the observation error in (c).

```
## $length
## [1] 500 500     1
names(ar1.level.latent_hyper.samples[[1]])
## [1] "hyperpar" "latent"     "logdens"
```

It is possible to look at time series plots of selected posterior x_t samples; for example, samples 500, 1000, 1500, and 2000 are shown in Figure 3.16.

Based on samples of size *n.samp* = *2000* we can construct histograms at selected times, say at $t = 250$ and time $t = 500$, as shown in Figure 3.17.

```
latent.sample.time <- ar1.level.latent_hyper.samples %>%
  sapply(function(x)
    x$latent[501:1000])
par(mfrow = c(1, 2))
hist(latent.sample.time[, 250],
     xlab = expression(x[250]),
     ylab = "frequency",
     main = "")
hist(latent.sample.time[, 500],
     xlab = expression(x[500]),
     ylab = "frequency",
     main = "")
```

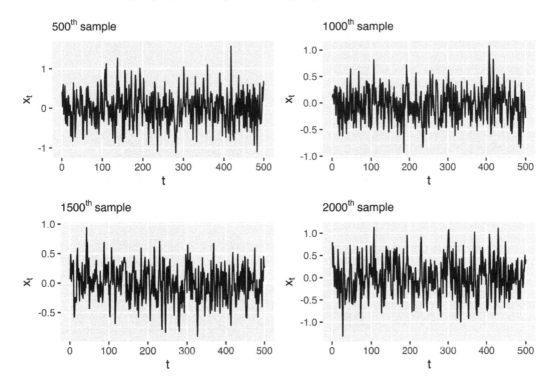

FIGURE 3.16: Time series plots of selected posterior samples from the AR(1) with level plus noise model.

We can obtain the predictive value $\hat{y}_t = E(y_t|y_1, \ldots, y_n)$ as follows:

```
ar1.level.latent_hyper.samples[[1]]$latent[1:n]
```

These can also be computed manually using posterior samples of x_t and α:

```
ar1.level.latent_hyper.samples[[1]]$latent[(n + 1):(2 * n)]
+ tail(ar1.level.latent_hyper.samples[[1]]$latent, 1)
```

While the values of $\hat{y}_t$ obtained from the two methods above are close, they do not coincide, as pointed out in Rue et al. (2009).

A note on reproducibility

R-INLA uses the OpenMP multiple processing interface. It uses multiple cores on shared memory by default; i.e., it uses all cores available in a machine for computation. Due to machine differences, small variations can be observed while running the same code on different machines and perhaps when repeating runs on any machine, depending on available cores. One can specify the number of cores by using the option `num.threads` in `inla()`, higher values offering speedup. Setting *num.threads = 1* can ensure exact reproducibility. See Wang et al. (2018) and R-INLA discussion group.[2]

---

[2]https://groups.google.com/g/r-inla-discussion-group/c/0dlvlVyiIRQ

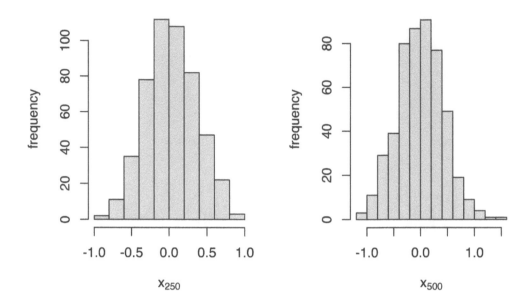

FIGURE 3.17: Histograms of posterior samples of the latent state for specific time points, t = 250 and t = 500, from the AR(1) model with level plus noise model.

3.11 Posterior predictive samples of unknown observations

In many time series examples, it is very useful to use the posterior samples of the latent effects and hyperparameters in order to generate samples from the *posterior predictive distributions* of unknown observations (responses) given the observed data. The function inla.posterior.sample.eval() can be used to do this, as shown in the following code.

The function inla.posterior.sample.eval calls the function fun.pred() in order to compute the posterior predictive samples of the holdout data (last $n_h = 5$ observations denoted as NA in the data frame).

```
fun.pred <- function(idx.pred) {
  m <- length(idx.pred)
  return (Predictor[idx.pred] + rnorm(m, mean = 0,
                          sd = sqrt(1 / theta[1])))
}
pred.post <-
  inla.posterior.sample.eval(fun.pred, ar1.level.latent_hyper.samples,
                      idx.pred = which(is.na(y.append)))
```

In Figure 3.18, we show plots of time series of the 500-th, 1000-th, 1500-th, and 2000-th posterior predictive samples for the holdout period, together with all the observed values of the time series.

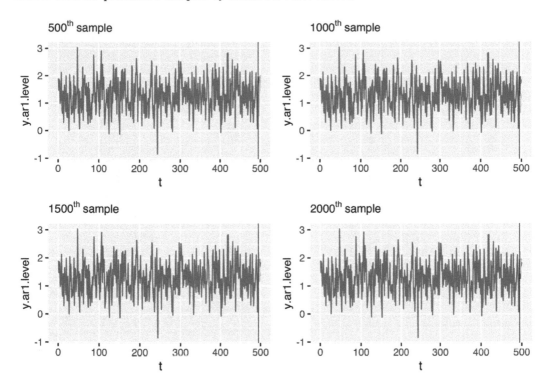

FIGURE 3.18: Comparing predictions (blue) of $n_h = 5$ holdout data and observed data (red). The black vertical line splits the data into the training and test sets.

We can also look at the histograms of the posterior predictive samples at times $t = 496$ and $t = 500$; see Figure 3.19.

```
par(mfrow = c(1, 2))
hist(pred.post[1, ],
     xlab = expression(x[496]),
     ylab = "frequency",
     main = "")
hist(pred.post[5, ],
     xlab = expression(x[500]),
     ylab = "frequency",
     main = "")
```

A comparison of the posterior means of the $n_h = 5$ holdout points obtained from `summary.fitted.values` and from the posterior predictive samples shows that they are very similar.

```
dim(pred.post)
## [1]    5 2000
tail(model.ar1.level$summary.fitted.values$mean, 5)
## [1] 1.526 1.409 1.583 1.549 1.547
unname(rowMeans(pred.post))
## [1] 1.542 1.407 1.602 1.569 1.549
```

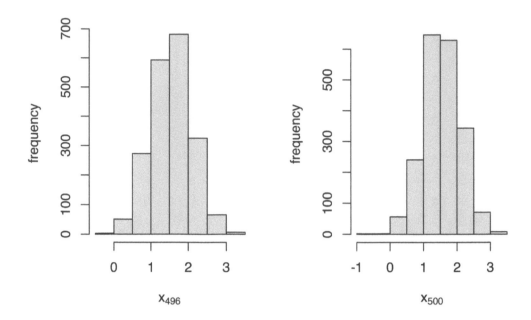

FIGURE 3.19: Histograms of posterior predictive samples for holdout time points t = 496 and t = 500 for the AR(1) with level plus noise model.

TABLE 3.1: Description of variables from inla() output.

Variable in INLA output	Math Notation	Description
Precision for the Gaussian observations	$1/\sigma_v^2$	Observation noise precision
Precision for id.w	$1/\sigma_w^2$	State noise precision
Precision for id.x	$1/\sigma_x^2$	State variable precision
Rho for id.x	ϕ	AR(1) coefficient
id.delta	δ	Drift in the state equation
(Intercept)	α	Level in the observation equation

We end this chapter with Table 3.1 that gives a quick reference for the correspondence between terms in R-INLA summaries in their output and the terms in the DLMs. As mentioned earlier, note that *Rho for id.x* denotes ϕ in the AR(1) model.

Chapter 3 – Appendix

Sampling properties of time series

Let X_t be a stationary time series.

1. Sample autocovariance function (ACVF)

The sample autocovariance function (ACVF) of X_t at lag h is

$$\hat{\gamma}(h) = \frac{1}{n} \sum_{t=1}^{n-|h|} (x_{t+|h|} - \bar{x})(x_t - \bar{x}), \quad -n < h < n. \tag{3.31}$$

Then, $\hat{\gamma}(0)$ is the sample variance of the time series.

2. Sample autocorrelation function (ACF)

The sample autocorrelation function (ACF) of X_t at lag h is

$$\hat{\rho}(h) = \hat{\gamma}(h)/\hat{\gamma}(0), \quad -n < h < n; \tag{3.32}$$

$\hat{\rho}(0) = 1$ and $|\hat{\rho}(h)| \le 1$, $h = \pm 1, \pm 2, \ldots$.

3. Sample spectrum-smoothed periodogram

Let $i = \sqrt{-1}$, and let $\omega_j = j/n$, for $j = 0, \ldots, n-1$ denote the Fourier frequencies. Define the complex-valued discrete Fourier transform (DFT) of X_t as

$$d(\omega_j) = \frac{1}{\sqrt{n}} \sum_{t=1}^{n} X_t \exp(-2\pi i \omega_j t). \tag{3.33}$$

The periodogram of X_t is the (real-valued) estimate of its spectral density function and is defined as

$$I(\omega_j) = |d(\omega_j)|^2. \tag{3.34}$$

The smoothed periodogram is a consistent estimate of the true spectrum of X_t and is obtained by averaging the periodogram ordinates over a frequency band of $L << n$ contiguous Fourier frequencies centered around ω_j. For example, the smoothed spectrum with the Daniell window is defined by

$$\hat{f}(\omega_j) = \frac{1}{L} \sum_{k=-m}^{m} I(\omega_j + \frac{k}{n}), \tag{3.35}$$

where, $L = 2m + 1$. The modified Daniell kernel is such that the two endpoints in the averaging get half the weight that the interior points get.

Autoregressive models

The AR(p) model is widely used in time series modeling and forecasting. Here, p is an integer which denotes the order of the model. The mean form of the AR(p) model is defined as

$$x_t - \mu = \phi_1(x_{t-1} - \mu) + \phi_2(x_{t-2} - \mu) + \ldots + \phi_p(x_{t-p} - \mu) + w_t, \tag{3.36}$$

while its intercept form is

$$x_t = \delta + \phi_1 x_{t-1} + \phi_2 x_{t-2} + \ldots + \phi_p x_{t-p} + w_t, \tag{3.37}$$

where $\delta = (1 - \phi_1 - \phi_2 - \ldots - \phi_p)\mu$. Using the backshift operator B, with $B^r x_t = x_{t-r}$, we can write the mean form and intercept form of the AR(p) model in operator notation as

$$(1 - \phi_1 B - \ldots - \phi_p B^p)(x_t - \mu) = w_t, \text{ or} \qquad (3.38)$$

$$(1 - \phi_1 B - \ldots - \phi_p B^p)x_t = \delta + w_t. \qquad (3.39)$$

The AR(p) process is stationary if the modulus of the roots of the AR polynomial equation

$$\phi(z) = 1 - \phi_1 z - \ldots - \phi_p z^p = 0 \qquad (3.40)$$

are all greater than 1. Recall that a polynomial equation of order p will have p roots. These roots may all be real, or some roots may be real, while other roots are conjugate complex. The modulus of a real root is its absolute value, while the modulus of a complex root $a + ib$ or $a - ib$ is $\sqrt{a^2 + b^2}$. We can use the R function `polyroot()` to find the roots of the AR polynomial equation, and verify whether the modulus of each root is greater than 1. In the `polyroot()` function, an ℓth order polynomial must be written in the form $p(z) = c_1 + c_2 z + \ldots + c_{\ell+1} z^\ell$, where the c_j, $j = 1, \ldots, \ell + 1$ are constants. For example, consider the AR(2) process

$$(1 - \phi_1 B - \phi_2 B^2)x_t = w_t,$$

with $\phi_1 = 1.5$ and $\phi_2 = -0.75$. The AR(2) polynomial is $p(z) = (1 - 1.5z + 0.75z^2)$, and the equation $p(z) = 0$ has two conjugate complex roots, i.e., $z[1] = 1 + 0.57735i$ and $z[2] = 1 - 0.57735i$, where $i = \sqrt{-1}$. The modulus of these roots is $1.1547 > 1$; hence this AR(2) process is stationary.

```
phi1 <- 1.5
phi2 <- -0.75
u <- polyroot(c(1,-phi1,-phi2))
u[1]
## [1] 1+0.577i
u[2]
## [1] 1-0.577i
Mod(u[1])
## [1] 1.155
Mod(u[2])
## [1] 1.155
```

The stationarity region for an AR(p) process is a convex set $C_p \subset \mathbb{R}^p$, and involves nonlinear constraints on $\phi = (\phi_1^p, \ldots, \phi_p^p)'$ To handle this, R-INLA uses the partial autocorrelation function (PACF) parameterization from C_p to the p-dimensional hypercube $(-1, 1)^p$ (see Barndorff-Nielsen and Schou (1973) and Marriott et al. (1996)). The one-to-one transformation with partial autocorrelations $r = (r_1, \ldots, r_p)'$ is

$$\phi_k^{(k)} = r_k$$

$$\phi_i^{(k)} = \phi_i^{(k-1)} - r_k \phi_{k-i}^{(k-1)}, \; i = 1, \ldots, k - 1.$$

Monahan (1984) gave the inverse transformation as

$$\phi_i^{(k-1)} = \frac{\phi_i^{(k)} + \phi_k^{(k)} \phi_{k-i}^{(k)}}{(1 - \phi_k^{(k)})^2}, \; i = 1, \ldots, k - 1, \qquad (3.41)$$

with Jacobian

$$J = \prod_{k=1}^{p}(1 - r_k^2)^{\frac{(k-1)}{2}} \prod_{j=1}^{[p/2]}(1 - r_{2j}). \tag{3.42}$$

In R-INLA, the partial autocorrelations of an $AR(p)$ model are denoted by ψ_k, with $|\psi_k| < 1$ for $k = 1, \ldots, p$. In terms of the internal parameterization, for $k = 1, \ldots, p$,

$$\psi_k = 2\frac{\exp(\theta_{\psi,k})}{1 + \exp(\theta_{\psi,k})} - 1.$$

For an $AR(2)$ process, these transformations are obtained as follows.

```
# phi to pacf
phi <- c(phi1, phi2)
pacf <- inla.ar.phi2pacf(phi)
#pacf to phi
phi.1 <- inla.ar.pacf2phi(pacf)
```

4

Modeling Univariate Time Series

4.1 Introduction

We present a real data illustration of a univariate time series using `R-INLA`. In Section 4.2, we describe the data (Musa, 1979). Sections 4.2.1–4.2.4 describe fitting different DLMs to the data. In Section 4.3, we obtain forecasts of future states and realizations. We then discuss model selection using both in-sample criteria based on adequacy of fit and out-of-sample criteria based on accuracy of forecasts. As we did in Chapter 3, our custom functions must be sourced so that they may be called throughout the chapter.

```
source("functions_custom.R")
```

The following `R` packages are used in this chapter.

```
library(tidyverse)
library(readxl)
library(INLA)
```

We analyze the following data set.

1. Musa data – a software engineering example

4.2 Example: A software engineering example – Musa data

During the software development phase, software goes through several stages of testing with the purpose of identifying and correcting "bugs" which cause failure of software. This process is referred to as debugging. Since it is expected that the debugging process improves the reliability of the software, such tests are known as reliability growth tests in the software engineering literature. Singpurwalla and Soyer (1992) consider modeling the time between failures in a software debugging process and discuss how the analysis of such data provides insights about assessment of *reliability growth*. The authors use *system 40* data from Musa (1979), which consists of observations on the times between failures over $n = 101$ testing stages.

We use a function from the `readxl` package to read the data from an *xlsx* file into `R`. Alternatively, one can also convert the data to a *csv* file and read it into R using the `read.csv()` function.

```
musa <- read_xlsx("MusaSystem40Data.xlsx")
head(musa)
```

```
## # A tibble: 6 x 1
##    `Time Beween Failures`
##                     <dbl>
## 1                     320
## 2                   14390
## 3                    9000
## 4                    2880
## 5                    5700
## 6                   21800
```

Let y_t denote the logarithms of the times between software failures, $t = 1, \ldots, 101$. Figure 4.1 shows a plot of y_t which exhibits an increasing trend over time, suggesting potential reliability growth.

```
musa <- musa %>%
  mutate(log.tbf = log(`Time Beween Failures`))
tsline(
  musa$log.tbf,
  xlab = "t",
  ylab = "log inter-failure time",
  line.size = 0.6,
  line.color = "red"
)
```

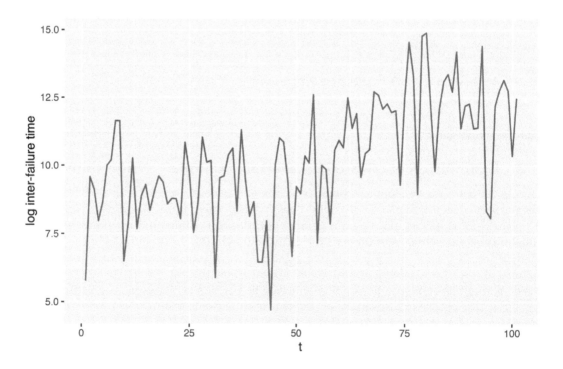

FIGURE 4.1: Logarithms of times between software failures.

We split the time series into a training (or calibration, or fitting) portion consisting of the first $n_t = 95$ observations, and a hold-out (or test, or validation) portion consisting of the last

$n_h = 6$ observations, with $n = n_t + n_h$. Under each model, we fit DLMs to the training data and use the fitted model to forecast the hold-out data. Recall that in R-INLA, there is no separate forecasting function like the `predict()` function in the `stats` package. Therefore, we append the end of the training data (of length $n_t = 95$) with NA's to represent the times to hold-out (the last $n_h = 6$ data points). The vectors *id* and *y.append* shown below each has length $n = 101$.

```
n <- nrow(musa)
n.hold <- 6
n.train <- n - n.hold
test.data <- tail(musa, n.hold)
train.data <- musa[1:n.train,]
y <- train.data$log.tbf
y.append <- c(y, rep(NA, n.hold))
```

Singpurwalla and Soyer (1992) point out that modifications made to the software at different stages of testing may potentially deteriorate the software's reliability even though the intention is to improve it. As a result, the expected time to failure of the software from one period to the next may not necessarily increase. This dynamic nature of the mean time to failure needs to be taken into account in modeling the behavior of times between software failures.

In Chapter 3, we described how to fit a DLM of the form

$$y_t = \alpha + x_t + v_t, \tag{4.1}$$
$$x_t = \phi x_{t-1} + w_t, \tag{4.2}$$

where, $\phi = 1\,(|\phi| < 1)$ gives us a random walk (AR(1)) model, and α denotes the level. In (4.2), the state variable x_t enables us to capture the dynamic behavior of the mean time to failure in the *Musa* data as a result of modifications made to the software at each stage. For example, when $\phi = 1$, the model implies that the mean is locally constant and the effect of modifications will be reflected by the magnitude of the state error variance. In other words, higher values of the state error variance will suggest a more drastic effect of modifications on the reliability of the software. By looking at the posterior distributions of x_t at each stage t, we can examine such behavior.

4.2.1 Model 1. AR(1) with level plus noise model

We fit an AR(1) with level plus noise model shown in (4.1) and (4.2) with $|\phi| < 1$ to the training data of length n_t.

```
id.x <- 1:length(y.append)
musa.ar1.dat <- cbind.data.frame(id.x, y.append)
```

Recall from Chapter 3 that *id.x* helps us to specify the random effect corresponding to the state variable x_t. The model formula and the `inla()` call are shown below.

```
formula.ar1 <- y.append ~ 1 + f(id.x, model = "ar1")
model.ar1 <- inla(
  formula.ar1,
  family = "gaussian",
  data = musa.ar1.dat,
  control.predictor = list(compute = TRUE),
```

```
  control.compute = list(
    dic = TRUE,
    waic = TRUE,
    cpo = TRUE,
    config = TRUE
  )
)
```

We show posterior summaries of the fixed and random parameters. The user may also obtain a complete summary using `summary(model.ar1)`.

```
# summary(model.ar1)
format.inla.out(model.ar1$summary.fixed[,c(1:5)])
##    name        mean   sd    0.025q 0.5q   0.975q
## 1 Intercept 10.02 0.869 8.155   10.03 11.78
format.inla.out(model.ar1$summary.hyperpar[,c(1,2,4,5)])
##    name                           mean   sd     0.5q   0.975q
## 1 Precision for Gaussian obs 0.416 0.070 0.411 0.567
## 2 Precision for id.x             0.713 0.423 0.621 1.783
## 3 Rho for id.x                   0.955 0.031 0.962 0.992
```

As we also discussed in Chapter 3, `Precision for id.x` value in the output does not refer to $1/\sigma_w^2$, but to $1/\sigma_x^2$, the precision of x_t in the stationary AR(1) process. To recover the posterior mean of the distribution of σ_w^2, we must use

```
sigma2.w.post.mean <- inla.emarginal(
  fun = function(x)
    exp(-x),
  marginal = model.ar1$internal.marginals.hyperpar$`Log precision for id.x`
) * (1 - model.ar1$summary.hyperpar["Rho for id.x", "mean"] ^ 2)
print(sigma2.w.post.mean)
## [1] 0.1736
```

In this model, the fitted values are the means of the smoothing distributions of x_t's and provide us with an overall assessment of modifications made to the software. The upward trend in the posterior means of x_t's suggests an overall improvement in software reliability as a result of the modifications. However, to assess the local effects from one stage to another, we must look at the means from the *filtering distributions* of x_t's. We compute the filtered means of x_t using code from the custom functions that we sourced, and then plot these along with observed y_t and smoothed estimates of x_t from Model 1; see Figure 4.2. We can recover the fitted values of y_t using the code

```
# model.ar1$summary.fitted.values
format.inla.out(head(model.ar1$summary.fitted.values))
##    name           mean   sd    0.025q 0.5q   0.975q mode
## 1 fit.Pred.001 8.711 0.749 7.133   8.748 10.09  8.824
## 2 fit.Pred.002 8.872 0.661 7.519   8.889 10.13  8.926
## 3 fit.Pred.003 8.953 0.621 7.695   8.965 10.15  8.989
## 4 fit.Pred.004 9.010 0.599 7.798   9.019 10.17  9.038
## 5 fit.Pred.005 9.149 0.582 7.995   9.149 10.30  9.150
## 6 fit.Pred.006 9.328 0.585 8.210   9.315 10.52  9.289
```

and the smooth estimates of x_t from the code

```
# model.ar1$summary.random$id.x
format.inla.out(head(model.ar1$summary.random$id.x))
##    ID mean    sd     0.025q 0.5q    0.975q mode   kld
## 1 1  -1.304 1.047 -3.483 -1.272 0.726  -1.201 0
## 2 2  -1.144 1.010 -3.226 -1.125 0.847  -1.084 0
## 3 3  -1.062 1.000 -3.114 -1.049 0.928  -1.021 0
## 4 4  -1.006 0.998 -3.045 -0.996 0.994  -0.975 0
## 5 5  -0.867 1.000 -2.899 -0.864 1.153  -0.858 0
## 6 6  -0.687 1.013 -2.734 -0.692 1.374  -0.707 0
```

As expected, the posterior means of the filtering distribution of the state variable, which capture the effects of the modifications at each time period, show more variation than the smoothing distribution means, and follow the local trends in the observed data more closely. Note that the gap between the posterior means of the filter estimates of x_t and the observed series is due to the presence of a level α in the model. The posterior mean of α is 10.015.

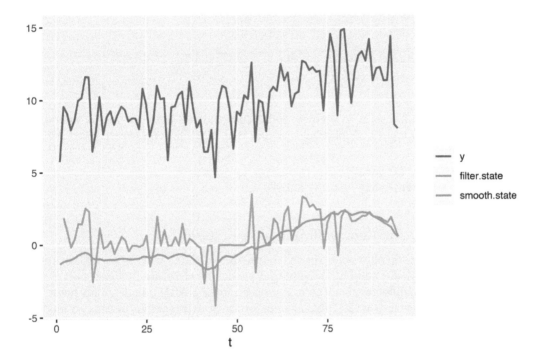

FIGURE 4.2: Observed data (red) and posterior means from filtered (yellow) and smoothed (green) distributions for the AR(1) with level plus noise model (Model 1).

The plot of the marginal posterior distribution of ϕ is shown in Figure 4.3.

```
plot(
  inla.smarginal(model.ar1$marginals.hyperpar$`Rho for id.x`),
  main = "",
  xlab = expression(phi),
  ylab = "density",
  type = "l"
)
```

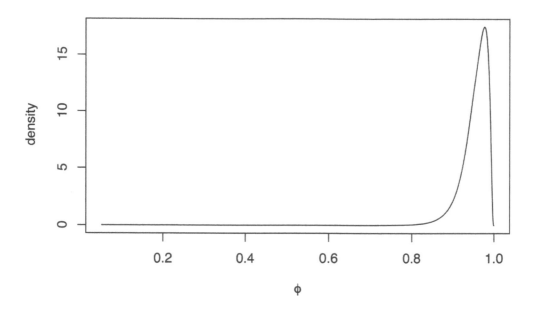

FIGURE 4.3: Marginal posterior density of ϕ in Model 1.

For each model, it is useful to recover and assess the residuals for model adequacy, by verifying that the sample ACF plot behaves like that of a white noise series; see the top left panel in Figure 4.9.

```
resid.ar1 <- y - fit.ar1
acf.ar1 <- acf(resid.ar1,plot = FALSE)
```

4.2.2 Model 2. Random walk plus noise model

We fit a random walk plus noise model to the *Musa* data, with observation and state equations given by (1.20) and (1.21). Note that we denote the state random effect by *id.w* in this model, which is different than the *id.x* we used for the AR(1) model. This helps us keep in mind the difference in what R-INLA outputs for the state precisions in the two models. We first fitted a model with level α; however, since the estimated α indicated that the level was close to 0, we show below results from a model without level α in the observation equation (by including −1 in the formula below).

```
id.w <- 1:length(y.append)
musa.rw1.dat <- cbind.data.frame(y.append, id.w)
formula.rw1 <- y.append ~ f(id.w, model = "rw1", constr = FALSE) - 1
model.rw1 <- inla(
  formula.rw1,
  family = "gaussian",
  data = musa.rw1.dat,
  control.predictor = list(compute = TRUE),
  control.compute = list(
    dic = TRUE,
```

```
    waic = TRUE,
    cpo = TRUE,
    config = TRUE
  )
)
summary.rw1 <- summary(model.rw1)
format.inla.out(model.rw1$summary.hyperpar[,c(1:2)])
##    name                        mean   sd
## 1 Precision for Gaussian obs  0.40   0.062
## 2 Precision for id.w          19.13  15.332
```

Note that in this case, the posterior mean of σ_w^2 can be obtained as 0.0844 using the code given below.

```
sigma2.w.post.mean.rw1 <- inla.emarginal(
  fun = function(x)
    exp(-x),
  marginal = model.rw1$internal.marginals.hyperpar$`Log precision for id.w`
)
```

Plots of the posterior means from the filtering and smoothing distributions from the random walk plus noise model (Model 2) are shown in Figure 4.4. In this model also, the fitted values of y_t correspond to the smooth estimates of x_t.

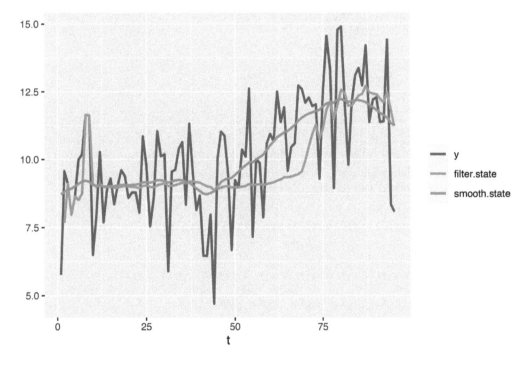

FIGURE 4.4: Observed data (red) and posterior means from the filtering (yellow) and smoothing (green) distributions for a random walk plus noise model (Model 2).

In Figure 4.1, the plot of the data showed an upward trend over time. As we usually do in time series analysis, we can include a term βt in the observation equation to account for this effect. This will be discussed next. In our development, we show the R-INLA model formulations, followed by plots of observed and fitted values, but suppress the detailed output. These can be seen in the GitHub page for our book.

4.2.3   Model 3. AR(1) with trend plus noise model

Previously, we fitted an AR(1) with level plus noise model (Model 1); now the model includes a time trend βt which is indicated by the fixed effect term *trend* in the **formula**. This model does not include an intercept α in the observation equation. The observation equation is

$$y_t = \beta t + x_t + v_t, \tag{4.3}$$

while the state equation is the same as in Model 1, i.e., (1.25). Since ϕ is close to 1, the user may fit a random walk with trend plus noise model instead of the AR(1) with trend plus noise model, and compare it with the other models.

```
id.x <- trend <- 1:length(y.append)
musa.ar1trend.dat <- cbind.data.frame(id.x, y.append, trend)
formula.ar1trend <- y.append ~ f(id.x, model = "ar1") + trend -1
model.ar1trend <- inla(
  formula.ar1trend,
  family = "gaussian",
  data = musa.ar1trend.dat,
  control.predictor = list(compute = TRUE),
  control.compute = list(
    dic = TRUE,
    waic = TRUE,
    cpo = TRUE,
    config = TRUE
  )
)
```

Partial output is shown below. Plots of the actual versus fitted values are shown in Figure 4.5. Unlike the random walk plus noise model, the fit does not show the smoothed x_t values, but rather the smoothed values plus the estimated trend.

```
# summary(model.ar1trend)
format.inla.out(model.ar1trend$summary.fixed[,c(1:5)])
##    name   mean   sd      0.025q  0.5q   0.975q
## 1 trend  0.074  0.043  -0.009   0.074  0.158
format.inla.out(model.ar1trend$summary.hyperpar[,c(1:2)])
##    name                         mean   sd
## 1 Precision for Gaussian obs   0.429  0.074
## 2 Precision for id.x           0.076  0.050
## 3 Rho for id.x                 0.992  0.008
```

The ACF plot of the residuals from Model 3 will be shown in Figure 4.9.

```
fit.ar1trend <- model.ar1trend$summary.fitted.values$mean[1:n.train]
resid.ar1trend <- y - fit.ar1trend
acf.ar1trend <- acf(resid.ar1trend, plot = FALSE)
```

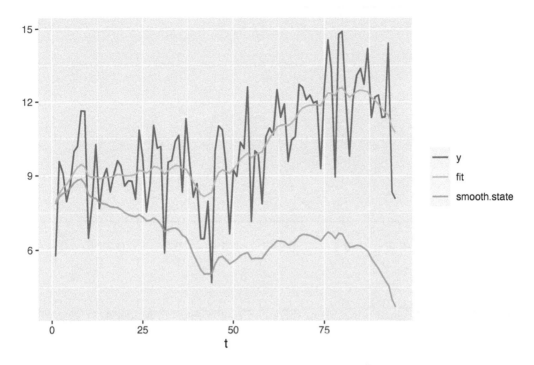

FIGURE 4.5: Observed (red) and fitted (blue) responses and posterior means from the smoothing distribution (green) for the AR(1) with trend plus noise model (Model 3).

We note that the posterior distribution of the trend coefficient β is concentrated around positive values (see Figure 4.6), implying the global upward trend in the data.

```
plot(
  inla.smarginal(model.ar1trend$marginals.fixed$trend),
  type = "l",
  main = "",
  xlab = expression(beta),
  ylab = "density"
)
```

4.2.4   Model 4. AR(2) with level plus noise model

We next consider an AR(2) with level plus noise model without a trend. The observation and state equations are given in (3.7) and (3.8), with $p = 2$, i.e.,

$$y_t = \alpha + x_t + v_t; \quad v_t \sim N(0, \sigma_v^2), \tag{4.4}$$

$$x_t = \sum_{j=1}^{2} \phi_j x_{t-j} + w_t; \quad w_t \sim N(0, \sigma_w^2). \tag{4.5}$$

Plots of the posterior means of x_t from the filtering and smoothing distributions for Model 4 are shown in Figure 4.7, together with the observed time series.

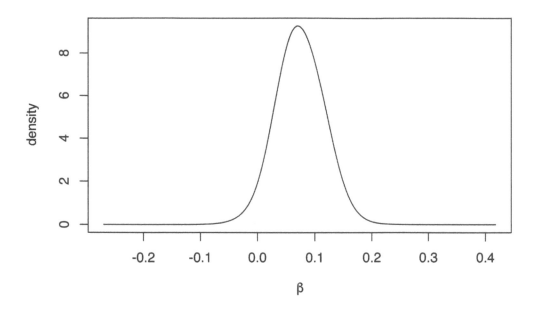

FIGURE 4.6: Marginal posterior of the trend coefficient β in Model 3.

```
id.x <- 1:length(y.append)
musa.ar2.dat <- cbind.data.frame(id.x, y.append)
formula.ar2 <- y.append ~ 1 + f(id.x,
                                    model = "ar",
                                    order = 2,
                                    constr = FALSE)
model.ar2 <- inla(
  formula.ar2,
  family = "gaussian",
  data = musa.ar2.dat,
  control.predictor = list(compute = TRUE),
  control.compute = list(
    dic = TRUE,
    waic = TRUE,
    cpo = TRUE,
    config = TRUE
  )
)
# summary(model.ar2)
format.inla.out(model.ar2$summary.fixed[,c(1:5)])
##   name       mean  sd     0.025q 0.5q  0.975q
## 1 Intercept 9.999 1.028 7.863  10.01 12.04
```

```
format.inla.out(model.ar2$summary.hyperpar[,c(1:2)])
##   name                              mean  sd
## 1 Precision for Gaussian obs 0.419 0.067
## 2 Precision for id.x             0.528 0.320
## 3 PACF1 for id.x                 0.960 0.027
## 4 PACF2 for id.x                 0.035 0.179
```

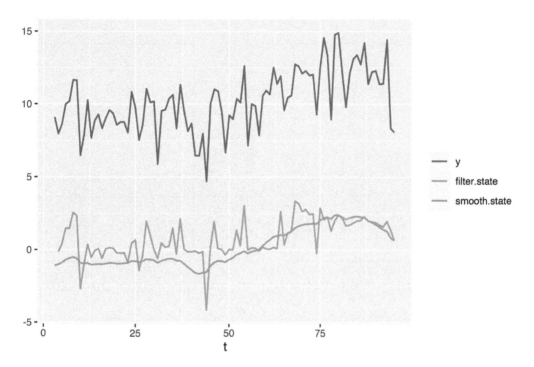

FIGURE 4.7: Observed data (red) and posterior means from the filtering (yellow) and smoothing (green) distributions for an AR(2) with level plus noise model (Model 4).

Plots of the marginal posterior distributions of the hyperparameters ϕ_1 and ϕ_2 are shown in Figure 4.8. The posterior mean of ϕ_2 is concentrated around 0, suggesting that the AR(1) model is more appropriate for this series.

```
n.samples <- 10000
pacfs <- inla.hyperpar.sample(n.samples, model.ar2)[, 3:4]
phis <- apply(pacfs, 1L, inla.ar.pacf2phi)
par(mfrow = c(1, 2))
plot(
  density(phis[1, ]),
  type = "l",
  main = "",
  xlab = expression(phi[1]),
  ylab = "density"
)
plot(
  density(phis[2, ]),
```

```
  type = "l",
  main = "",
  xlab = expression(phi[2]),
  ylab = "density"
)
```

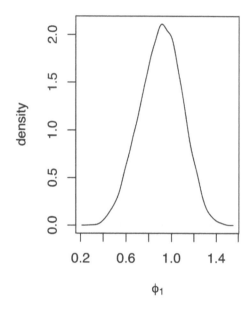 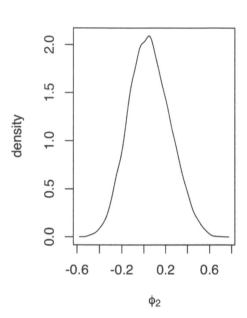

FIGURE 4.8: Marginal posterior densities of ϕ_1 (left panel) and ϕ_2 (right panel) under Model 4.

In Section 4.4, we will compare these models based on in-sample model criteria such as DIC, WAIC, and PsBF. In Section 4.3, we see how to use R-INLA to forecast future states and observations.

4.3   Forecasting future states and responses

Given training data $y_1, \ldots, y_{n_t}$, estimates of the states $x_1, \ldots, x_{n_t}$ and the estimates of model hyperparameters $\boldsymbol{\theta}$, we forecast $x_{n_t+\ell}$, $\ell = 1, \ldots, n_h$. For $n_t = 95$ and $n_h = 6$, we show below forecasts for the future latent states for time periods from $t = 96$ to $t = 101$. See Section 3.7 for more details.

```
# Forecast the state variable
format.inla.out(tail(model.ar1$summary.random$id[,c(1:6)]))
##    ID  mean   sd     0.025q  0.5q   0.975q
## 1  96 0.568  1.132  -1.704  0.572  2.878
```

Model 1: AR(1) with level plus noise

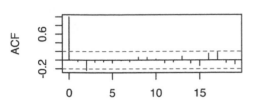

Model 2: RW plus noise

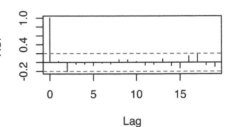

Model 3: AR(1) with trend plus noise

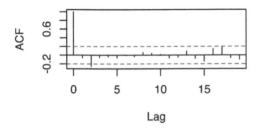

Model 4: AR(2) with level plus noise

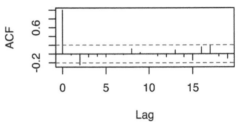

FIGURE 4.9: Sample ACF plots from Model 1–Model 4.

```
## 2   97 0.547 1.166 -1.809 0.555 2.910
## 3   98 0.527 1.193 -1.893 0.537 2.938
## 4   99 0.508 1.217 -1.964 0.518 2.963
## 5  100 0.490 1.236 -2.023 0.500 2.984
## 6  101 0.472 1.253 -2.074 0.482 3.001
```

We then obtain forecasts of $y_{n_t+\ell}$ for $\ell = 1, \ldots, n_h$.

```
format.inla.out(tail(model.ar1$summary.fitted.values[,c(1:5)]))
##   name          mean  sd     0.025q 0.5q   0.975q
## 1 fit.Pred.096 10.58 0.891 8.664   10.64 12.18
## 2 fit.Pred.097 10.56 0.955 8.493   10.63 12.27
## 3 fit.Pred.098 10.54 1.009 8.358   10.61 12.35
## 4 fit.Pred.099 10.52 1.055 8.245   10.59 12.43
## 5 fit.Pred.100 10.51 1.094 8.149   10.57 12.49
## 6 fit.Pred.101 10.49 1.128 8.067   10.55 12.55
```

Using similar code, we can obtain forecasts for $y_{n_t+1}, \ldots, y_{n_t+n_h}$ from Model 2–Model 4. Figure 4.10 shows plots of the training portion of the time series as well as the forecasts and forecast intervals (gray shaded region) from all four models. The shaded area is the interval at each time point between the lower limit (forecast minus twice the forecast standard deviation) and the upper limit (forecast plus twice the forecast standard deviation).

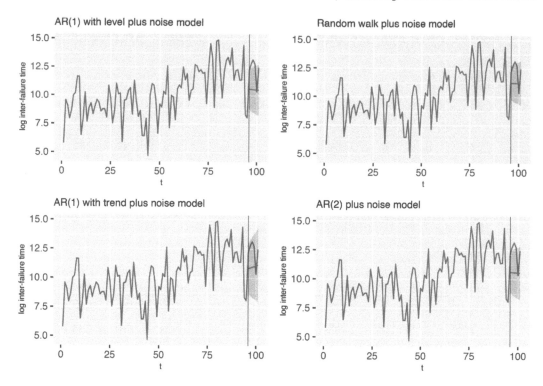

FIGURE 4.10: Observed data along with forecasts from Model 1–Model 4 for the last six observations. The gray shaded region shows the forecast intervals based on 2 standard deviation (sd) bounds around the means of the posterior predictive distributions. The black vertical line splits the data into training and test portions.

4.4 Model comparisons

In the next two sections, we present model comparisons.

In-sample comparisons

The DIC, WAIC, and CPO criteria for the fitted models are obtained using `control.compute=list(dic=TRUE, waic = TRUE, cpo = TRUE)` during model fit:

```
model.ar1$dic$dic
## [1] 368.9
model.ar1$waic$waic
## [1] 371
cpo.ar1 <- model.ar1$cpo$cpo
(PsBF.ar1 <- sum(log(cpo.ar1[1:n.train])))
## [1] -186.1
```

The custom function `model.selection.criteria()` (see Section 14.3) allows us to print the DIC, WAIC and PSBF values from a model. For example, we can obtain these for the AR(1) with level plus noise model (Model 1) as shown below.

TABLE 4.1: DIC, MAPE, and MAE values from Models M1–M4.

Model	DIC	MAPE	MAE
AR(1) with level plus noise	368.9	14.25	1.807
Random walk plus noise	370.1	10.82	1.344
AR(1) with trend (no level alpha) and noise	368.5	12.40	1.551
AR(2) plus noise	368.7	14.04	1.778

```
model.selection.criteria(model.ar1, plot.PIT = FALSE, n.train = n.train)
##         DIC WAIC    PsBF
## [1,] 368.9   371  -186.1
```

R-INLA outputs the marginal likelihood on the logarithmic scale by default; for example the marginal likelihood from Model 1 is:

```
model.ar1$mlik
##                                             [,1]
## log marginal-likelihood (integration) -211.2
## log marginal-likelihood (Gaussian)    -211.2
```

From Table 4.1, we see that the criteria are quite similar for four models. The DIC criterion suggests that either the AR(1) with trend plus noise model seems to give the best fit to the data, while the other models are also competitive.

Out-of-sample comparisons

We defined out-of-sample criteria such as MAPE and MAE in Chapter 3; see (3.28) and (3.29). The codes to compute MAE and MAPE can be accessed from the custom functions; see Section 14.3. The out-of-sample validation based on forecasting the hold-out data for the AR(1) with level plus noise model (Model 1) is shown below. These criteria can be similarly computed for the other models.

```
yfore.ar1 <- tail(model.ar1$summary.fitted.values)$mean
yhold <- test.data$log.tbf
efore.ar1 <- yhold - yfore.ar1 # forecast errors
mape(yhold, yfore.ar1, "MAPE from AR(1) with level plus noise model is")
## [1] "MAPE from AR(1) with level plus noise model is 14.25"
mae(yhold, yfore.ar1, "MAE from AR(1) with level plus noise model is")
## [1] "MAE from AR(1) with level plus noise model is 1.807"
```

The MAPE and MAE values from Table 4.1 are lowest for the random walk plus noise model (Model 2), with the AR(1) with trend plus noise model being a close second.

5

Time Series Regression Models

5.1 Introduction

This chapter discusses time series regression models. The response is a univariate time series that we wish to model and forecast. We discuss two types of predictors: (a) predictors that we construct as functions of time and include in a structural decomposition model, and (b) exogenous time series that are observed on the same time frame and may be associated with the response time series at contemporaneous or lagged times. We also show how we can build a latent AR(1) model with fixed, exogenous predictors.

As in previous chapters, the custom functions must be sourced.

```
source("functions_custom.R")
```

The list of packages used in this chapter are shown below.

```
library(INLA)
library(astsa)
library(lubridate)
library(tidyverse)
library(gridExtra)
```

We show analyses for the following data.

1. Monthly average cost of nightly hotel stay

2. Hourly traffic volumes

5.2 Structural models

This section casts the well known structural decomposition of a univariate time series as a dynamic linear model (DLM) and shows how to fit and forecast the time series using R-INLA. An observed time series may exhibit features such as trend and seasonality, in addition to stochastic dependence. To build a structural time series model (also called a *classical decomposition model* in the literature), we decompose a time series y_t as the sum of different components:

$$y_t = \text{Level} + \text{Trend} + \text{Seasonal Component} + \text{Random Component}, \; i.e.,$$
$$y_t = \alpha + m_t + S_t + v_t, \tag{5.1}$$

where α is a level coefficient, m_t denotes a latent trend component, S_t denotes a latent seasonal component, and v_t denotes a random error component. The seasonal period is an integer such as $s = 4, 7$, or 12, corresponding to time series that exhibits a quarterly, daily, or monthly effect. The structural decomposition models have wide application in business, economics, etc.; see Harvey and Koopman (2014).

5.2.1   Example: Monthly average cost of nightly hotel stay

The data consists of the monthly average cost of a night's accommodation at a hotel, motel or guesthouse in Victoria, Melbourne, from January 1980 to June 1995. In Figure 5.1, we show plots of the time series (cost), as well as the natural logarithm of the series (lcost). The plot of the nonstationary time series *lcost* stabilizes the variance in *cost*, and exhibits an increasing trend.

```
hotels <- read.csv("NightlyHotel.csv")
str(hotels)
## 'data.frame':    186 obs. of  3 variables:
##  $ Yr_mth: chr  "1980-01" "1980-02" "1980-03" "1980-04" ...
##  $ Cost  : num  27.7 28.7 28.6 28.3 28.6 ...
##  $ CPI   : num  45.2 45.2 45.2 46.6 46.6 46.6 47.5 47.5 47.5 48.5 ...
cost <- ts(hotels[, 2])
lcost <- log(cost)
c1 <-
  tsline(
    as.numeric(cost),
    xlab = "Month",
    ylab = "cost",
    line.size = 0.6,
    line.color = "black"
  )
lc1 <-
  tsline(
    as.numeric(lcost),
    xlab = "Month",
    ylab = "log cost",
    line.size = 0.6,
    line.color = "red"
  )
grid.arrange(c1, lc1, nrow = 1)
```

We fit a DLM to the nonstationary series *lcost* to handle the time-varying trend with additive seasonality (with $s = 12$), deferring a discussion of conditional heteroscedasticity until Chapter 10. Keeping in mind that R-INLA has no predict function, we must append the training data with NA's to represent the hold-out times. We hold out six months of data for predictive cross validation. Each of the vectors *trend* and *y.append* has length $n = 186$, with $n_t = 180$ and $n_h = 6$ denoting the training and hold-out samples respectively. We bind these columns to create a data frame.

```
n <- length(lcost)
n.hold <- 6
n.train <- n - n.hold
test.lcost <- tail(lcost, n.hold)
```

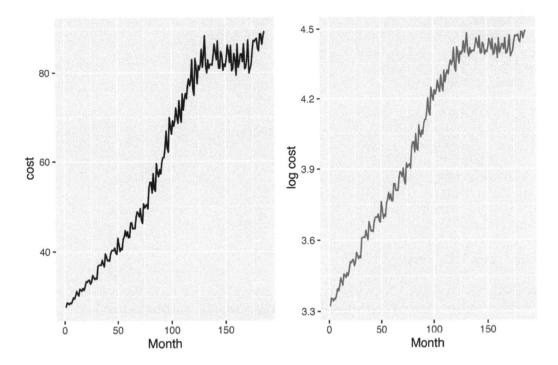

FIGURE 5.1: Average monthly cost (left plot) and log cost (right plot).

```
train.lcost <- lcost[1:n.train]
y <- train.lcost
y.append <- c(y, rep(NA, n.hold))
n.seas <- 12
trend <- 1:length(y.append)
lcost.dat <- cbind.data.frame(y = y.append,
                              trend = trend,
                              seasonal = trend)
```

We model the latent trend m_t by a random walk

$$m_t = m_{t-1} + w_{t,1}, \tag{5.2}$$

where $w_{t,1}$ is a Gaussian state error process. The seasonal variation in *lcost* with periodicity $s = 12$ can be modeled by assuming that $\sum_{j=0}^{s-1} S_{t-j}$ are independent Gaussian random variables, i.e.,

$$S_t = -\sum_{j=1}^{11} S_{t-j} + w_{t,2}. \tag{5.3}$$

For details, see `inla.doc("seasonal")`. Equations (5.2) and (5.3) are the state equations in the DLM for *lcost*, while the observation equation is given by (5.1). When we fitted a model with level α, we saw that this coefficient was not significant. We therefore show the code and results for a model without level α.

```
formula.lcost.rw1 <- y.append ~ -1 +
  f(trend, model = "rw1", constr = FALSE) +
  f(seasonal, model = "seasonal",
    season.length = n.seas)
model.lcost.rw1 <- inla(
  formula.lcost.rw1,
  family = "gaussian",
  data = lcost.dat,
  control.predictor = list(compute = TRUE),
  control.compute = list(
    dic = TRUE,
    waic = TRUE,
    cpo = TRUE,
    config = TRUE
  )
)
summary(model.lcost.rw1)
##
## Call:
##    c("inla(formula = formula.lcost.rw1, family = \"gaussian\",
##    data = lcost.dat, ", " control.compute = list(dic = TRUE, waic
##    = TRUE, cpo = TRUE, ", " config = TRUE), control.predictor =
##    list(compute = TRUE))" )
## Time used:
##    Pre = 4.68, Running = 0.302, Post = 0.248, Total = 5.23
## Random effects:
##   Name      Model
##     trend RW1 model
##     seasonal Seasonal model
##
## Model hyperparameters:
##                                              mean        sd 0.025quant
## Precision for the Gaussian observations 38811.25 21587.79   12937.41
## Precision for trend                      3937.54   636.43    2821.77
## Precision for seasonal                  55786.67 18354.44   28280.75
##                                         0.5quant 0.975quant     mode
## Precision for the Gaussian observations 33678.14   94599.50 25854.69
## Precision for trend                      3891.96    5320.38  3806.62
## Precision for seasonal                  52980.47   99654.59 47811.14
##
## Expected number of effective parameters(stdev): 157.84(9.71)
## Number of equivalent replicates : 1.14
##
## Deviance Information Criterion (DIC) ..............: -1202.97
## Deviance Information Criterion (DIC, saturated) ....: 334.93
## Effective number of parameters ....................: 156.14
##
## Watanabe-Akaike information criterion (WAIC) ...: -1227.92
## Effective number of parameters .................: 98.53
##
```

```
## Marginal log-Likelihood:   387.22
## CPO and PIT are computed
##
## Posterior marginals for the linear predictor and
##  the fitted values are computed
```

To see all the items in the list summary(model.lcost.rw1), we can use

```
names(summary(model.lcost.rw1))
##  [1] "call"             "cpu.used"          "hyperpar"
##  [4] "random.names"     "random.model"      "neffp"
##  [7] "dic"              "waic"              "mlik"
## [10] "cpo"              "linear.predictor"  "family"
```

The condensed output for the random effects is shown below.

```
format.inla.out(model.lcost.rw1$summary.hyperpar[,c(1:2)])

##   name                      mean  sd
## 1 Precision for Gaussian obs 38811 21587.8
## 2 Precision for trend         3938   636.4
## 3 Precision for seasonal     55787 18354.4
```

Plots of the posterior means of m_t and S_t along with the sample spectrum, ACF, and PACF of S_t in Figure 5.2 show evidence of periodic behavior with $s = 12$.

The marginal likelihood, DIC, and WAIC are shown below:

```
summary(model.lcost.rw1)$mlik
##                                          [,1]
## log marginal-likelihood (integration) 387.3
## log marginal-likelihood (Gaussian)    387.2
summary(model.lcost.rw1)$dic$dic
## [1] -1203
summary(model.lcost.rw1)$waic$waic
## [1] -1228
```

We can obtain forecasts for future state variables, $m_{n_t+\ell}$ and $S_{n_t+\ell}$, for $\ell = 1, \ldots, n_h$.

```
format.inla.out(tail(model.lcost.rw1$summary.random$trend, n.hold)[,c(2:6)])
##   name mean  sd     0.025q 0.5q   0.975q
## 1 181  4.465 0.019 4.428  4.465  4.502
## 2 182  4.465 0.025 4.416  4.465  4.513
## 3 183  4.465 0.029 4.407  4.465  4.523
## 4 184  4.465 0.034 4.399  4.465  4.531
## 5 185  4.465 0.037 4.392  4.465  4.538
## 6 186  4.465 0.041 4.385  4.465  4.545
format.inla.out(tail(model.lcost.rw1$summary.random$seasonal,
                     n.hold)[,c(2:5)])
##   name mean   sd    0.025q 0.5q
## 1 181  -0.027 0.01 -0.046 -0.027
## 2 182   0.008 0.01 -0.012  0.008
## 3 183   0.024 0.01  0.004  0.024
## 4 184  -0.037 0.01 -0.057 -0.037
```

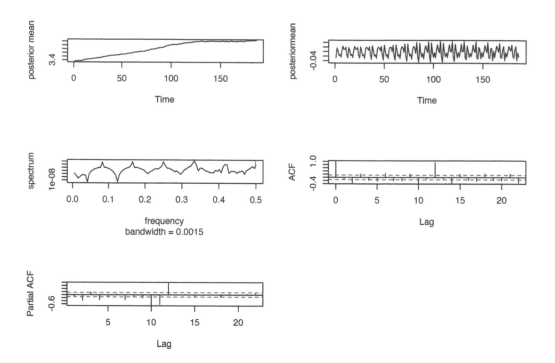

FIGURE 5.2: The top plot (left panel) shows the posterior means of m_t, while the top plot (right panel) is the posterior mean of S_t. The plot in the middle (left panel) is the spectrum of the posterior means of S_t, the plot in the middle (right panel) is the ACF of the posterior means of S_t, and the plot in the bottom (left panel) is the PACF of S_t.

```
## 5 185   -0.003 0.01 -0.023 -0.003
## 6 186   -0.021 0.01 -0.040 -0.021
```

Forecasts of $y_{n_t+\ell}$ for $\ell = 1, \ldots, n_h$ can also be obtained.

```
format.inla.out(tail(model.lcost.rw1$summary.fitted.values,n.hold)[,c(1:5)])
##    name           mean   sd    0.025q 0.5q  0.975q
## 1 fit.Pred.181  4.438  0.022  4.395  4.438 4.482
## 2 fit.Pred.182  4.473  0.027  4.420  4.473 4.526
## 3 fit.Pred.183  4.489  0.031  4.427  4.489 4.551
## 4 fit.Pred.184  4.427  0.035  4.358  4.427 4.497
## 5 fit.Pred.185  4.461  0.039  4.385  4.461 4.538
## 6 fit.Pred.186  4.444  0.042  4.362  4.444 4.527
```

The MAPE and MAE values corresponding to these forecasts are 0.441 and 0.02 respectively. The fitted *lcost* time series shown in the left panel of Figure 5.3 indicates a very close fit with the observed series. Both series show a leveling off pattern. The plot also shows the observations and forecasts for the $n_h = 6$ values of the hold-out sample of *lcost*. There is some disparity since the observed *lcost* appears to be increasing in the test portion, while the forecast values continue the leveling off pattern. The right panel shows the corresponding plots for the *cost* series.

```
fit.lcost.rw1 <- model.lcost.rw1$summary.fitted.values$mean
plot.data.rw1 <-
  as_tibble(cbind.data.frame(
    time = trend,
    lcost = lcost,
    lcost.fit = fit.lcost.rw1
  ))
p.lcost <- multiline.plot(
  plot.data.rw1,
  xlab = "Month",
  ylab = "",
  line.type = "solid",
  line.size = 0.6,
  line.color = c("red", "blue")
) +
  geom_vline(xintercept = n.train, size = 0.2)
# cost
fit.cost <- exp(fit.lcost.rw1)
plot.data <-
  as_tibble(cbind.data.frame(
    time = trend,
    cost = cost,
    cost.fit = fit.cost
  ))
p.cost <- multiline.plot(
  plot.data,
  xlab = "Month",
  ylab = "",
  line.type = "solid",
  line.size = 0.6,
  line.color = c("red", "blue")
) +
  geom_vline(xintercept = n.train, size = 0.2)
grid.arrange(p.lcost, p.cost, layout_matrix = cbind(1,2))
```

5.3   Models with exogenous predictors

In many situations, we wish to model a time series y_t as a function of exogenous predictors in addition to functions of time to reflect trend or seasonality. The observation equation of the DLM has k exogenous variables $Z_1, \ldots, Z_k$ observed over time in addition to a latent state variable x_t and additive noise v_t. We treat the regression coefficients $\beta_1, \ldots, \beta_k$ corresponding to the exogenous variables to be fixed unknown constants. We can assume that the state equation has an $AR(p)$ or random walk structure; here, we show an $AR(1)$ state equation.

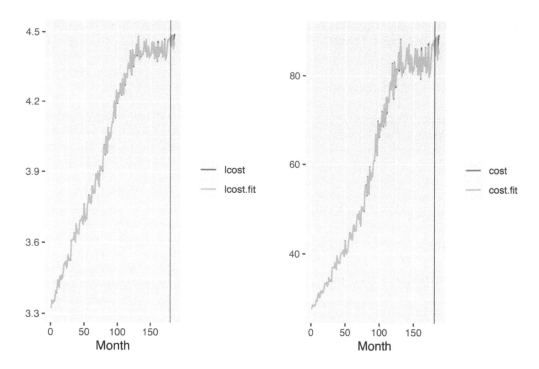

FIGURE 5.3: Observed (red) versus fitted (blue) lcost (left panel) and cost (right panel) from Model 1. The black vertical line divides the data into training and test portions in each plot.

$$y_t = x_t + \sum_{j=1}^{k} \beta_j Z_{t,j} + v_t; \quad v_t \sim N(0, \sigma_v^2), \tag{5.4}$$

$$x_t = \phi x_{t-1} + w_t; \quad w_t \sim N(0, \sigma_w^2), t = 2, \dots, n; \tag{5.5}$$

x_1 follows the normal distribution in (3.3), $|\phi| < 1$, $Z_{t,j}$, $j = 1, \dots, k$ are exogenous time series, and β_j, $j = 1, \dots, k$ are the corresponding regression coefficients.

5.3.1   Example: Hourly traffic volumes

We model hourly traffic volumes data by combining the framework in (5.1) with the equations (5.4)-(5.5). The data is available from the Center for Machine Learning and Intelligent Systems at the University of California website.[1] The hourly Interstate 94 Westbound traffic volume was recorded at the Automated Traffic Recording station 301 of Minnesota's Department of Transportation(DoT) located roughly midway between Minneapolis and St. Paul.

```
traffic <-
  read.csv(file = "Metro_Interstate_94 Traffic_Volume_Hourly.csv",
           header = TRUE,
           stringsAsFactors = FALSE)
traffic <- as_tibble(traffic)
```

---

[1]https://archive.ics.uci.edu/ml/datasets/Metro+Interstate+Traffic+Volume

```
traffic <- traffic %>%
  mutate(date_time = as.POSIXct(
    as.character(date_time),
    origin = "1899-12-30",
    format = "%m/%d/%Y %H",
    tz = "UTC"
  ))
traffic <- traffic %>%
  mutate(
    month = month(date_time, label = TRUE),
    day = day(date_time),
    year = year(date_time),
    dayofweeek = weekdays(date_time),
    hourofday = hour(date_time),
    weeknum = week(date_time)
  )
```

We fit models to three months of hourly data (in 1000's), consisting of $n = 2160$ observations. In Figure 5.4, we show plots of the time series (after normalizing the values by dividing them by 1000), the sample ACF and sample spectrum plotted versus frequency, as well as the amplitudes of the spectrum plotted versus period.

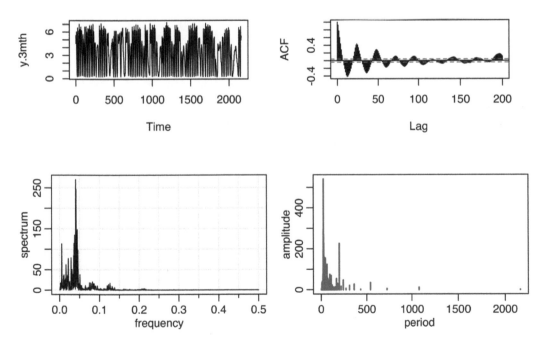

FIGURE 5.4: Time series plot (left) and ACF (right) of hourly traffic volume in the top panel and sample spectrum (left) and amplitude versus period (right) in the bottom panel.

We can identify the periods (or harmonics) associated with the highest periodogram amplitudes. Since this is hourly data, we will expect to see a period of 24. From the periodogram, we can see that the highest amplitude is indeed associated with period 24 (corresponding to harmonic 90) and values near 24. Periods in the vicinity of $7 \times 24 = 168$ point to daily periodicity. The sample periodogram points to harmonics in the vicinity of the periods 168 and 196 as having higher amplitudes. In addition, we can include harmonics with high amplitudes associated with other periods, and create sine and cosine terms associated with these periods that serve as predictors in the DLM.

```
id <- 1:n
k <- c(90, 11, 20, 47, 62, 23, 9, 34, 29, 13)
sin.mat <- cos.mat <- matrix(0, nrow = n, ncol = length(k))
for (i in 1:length(k)) {
  sin.mat[, i] <- sin(2 * pi * id * k[i] / n)
  colnames(sin.mat) <- paste("S", c(1:(length(k))), sep = "")
  cos.mat[, i] <- cos(2 * pi * id * k[i] / n)
  colnames(cos.mat) <- paste("C", c(1:(length(k))), sep = "")
}
```

It is also useful to see whether exogenous predictors relating to holiday and weather effects affect the traffic volume.

```
holiday <- traffic$holiday[1:n]
temp <- traffic$temp[1:n]
rain <- traffic$rain_1h[1:n]
snow <- traffic$snow_1h[1:n]
clouds <- traffic$clouds_all[1:n]
weather <- traffic$weather_main[1:n]
dayofweek <- traffic$dayofweeek[1:n]
hourofday <- traffic$hourofday[1:n]
```

We set up a data frame with all the variables.

```
trend <- 1:n
y.traffic <-
  cbind.data.frame(
    y.3mth, id, trend = trend, sin.mat, cos.mat,
    holiday, temp, rain, snow, clouds, weather,
    dayofweek, hourofday
  )
```

Model E1

We set up and fit a DLM with the leading harmonics as exogenous fixed predictors, and an AR(1) trend. Since we anticipate that the leading harmonics that we obtained empirically will capture significant *hourofday* and *dayofweek* effects, we exclude these two variables from the model.

```
formula.3mth.1 <-
  y.3mth ~ 1 + sin.mat +cos.mat +
  f(trend, model = "ar1", constr = FALSE)

model.3mth.1 <- inla(
  formula.3mth.1,
  family = "gaussian",
  data = y.traffic,
```

```
  control.predictor = list(compute = TRUE),
  control.compute = list(dic = TRUE, cpo = TRUE, waic = TRUE)
)
#summary(model.3mth.1)
```

The condensed summaries on estimated effects are shown below. If the interval (0.025q, 0.975q) does not include 0, we say that the effect is significant in the Bayesian sense (Berry, 1996).

```
format.inla.out(model.3mth.1$summary.fixed[,c(1:5)])
##      name         mean    sd     0.025q  0.5q    0.975q
## 1   Intercept    3.274  0.210   2.861   3.273  3.687
## 2   S1           0.447  0.088   0.274   0.447  0.620
## 3   S2          -0.132  0.276  -0.676  -0.132  0.412
## 4   S3           0.138  0.241  -0.336   0.138  0.610
## 5   S4           0.364  0.151   0.067   0.364  0.661
## 6   S5           0.304  0.122   0.065   0.304  0.543
## 7   S6           0.122  0.228  -0.327   0.122  0.570
## 8   S7           0.232  0.283  -0.326   0.232  0.788
## 9   S8          -0.104  0.188  -0.472  -0.104  0.265
## 10  S9           0.096  0.205  -0.307   0.096  0.499
## 11  S10          0.163  0.269  -0.367   0.163  0.692
## 12  C1           0.510  0.088   0.337   0.510  0.683
## 13  C2           0.448  0.274  -0.091   0.448  0.987
## 14  C3          -0.207  0.239  -0.677  -0.207  0.263
## 15  C4          -0.114  0.151  -0.410  -0.114  0.182
## 16  C5           0.241  0.122   0.002   0.241  0.479
## 17  C6           0.249  0.227  -0.198   0.249  0.695
## 18  C7           0.031  0.280  -0.521   0.031  0.583
## 19  C8          -0.314  0.187  -0.681  -0.314  0.053
## 20  C9           0.207  0.204  -0.194   0.207  0.608
## 21  C10          0.184  0.267  -0.341   0.184  0.709
format.inla.out(model.3mth.1$summary.hyperpar[,c(1:2)])
##    name                           mean         sd
## 1 Precision for Gaussian obs 16167.684  1.05e+04
## 2 Precision for trend             0.259  3.00e-02
## 3 Rho for trend                   0.922  9.00e-03
model.3mth.1$mlik
##                                              [,1]
## log marginal-likelihood (integration) -2585
## log marginal-likelihood (Gaussian)     -2587
```

We note that the cosine/sine pair (C1, S1) corresponding to the period 24 definitely contribute to the model. Also significant are S4 and S5 which are associated with (approximately) periods of 46 and 35, which in turn may be related to 2 day and 1.5 day cycles, respectively. The posterior mean of ϕ_1 is 0.98. We can also refit the model including the weather related variables; we do not show details here.

TABLE 5.1: In-sample model comparisons for the hourly traffic data.

Model	DIC	WAIC	PsBF
Model E1	−12989.964	−13432.837	4776.67
Model E2	−12876.462	−13294.348	4712.426

Model E2

In Model E2, we include *hourofday* and *dayofweek effects*, as well as weather/holiday variables as fixed exogenous predictors, together with an AR(1) trend, but exclude the empirically determined leading harmonics.

```
formula.3mth.2 <- y.3mth ~ 1 + hourofday + dayofweek +
  holiday + temp + rain + snow + clouds + weather +
  f(trend, model = "ar1", constr = FALSE)
model.3mth.2 <- inla(
  formula.3mth.2,
  family = "gaussian",
  data = y.traffic,
  control.predictor = list(compute = TRUE),
  control.compute = list(
    dic = TRUE,
    cpo = TRUE,
    waic = TRUE,
    config = TRUE
  )
)
format.inla.out(model.3mth.2$summary.hyperpar[,c(1:2)])
##   name                        mean        sd
## 1 Precision for Gaussian obs 21950.231 19576.110
## 2 Precision for trend             0.268     0.028
## 3 Rho for trend                   0.919     0.009
model.3mth.2$mlik   # marginal likelihood
##                                              [,1]
## log marginal-likelihood (integration) -2606
## log marginal-likelihood (Gaussian)     -2606
```

The DIC, WAIC, and PsBF values from the two models are displayed in Table 5.1. All these criteria prefer Model E2. The setup shown above can be used to investigate other models for the hourly traffic in order to identify the smallest significant model, but the details are not shown here.

Remark. Frequently, we must model time series exhibiting multiple seasonal periods. For example, the *soi* data in the R package **astsa** exhibits an annual cycle ($s_1 = 12$) and a four-year cycle ($s_2 = 48$). Call center volume data often exhibit an hour of day effect as well as a weekly, or monthly cycle. In a non-dynamic setting, we can use the **msts()** function in the R package **forecast** (Hyndman et al. (2021) and Hyndman and Khandakar (2008)).

Suppose s_1 and s_2 are two distinct seasonal periods. We can construct fundamental harmonics for each period, and include linearly independent columns as exogenous predictors in the DLM. We show three alternative model setups below.

1. We can include them as static predictors (together with a dynamic trend, if relevant), similar to Model E1. Suppose $s_1 = 12$ and $s_2 = 48$; then it is sufficient to include the harmonics for $s_2 = 48$ as predictors, since they include the harmonics for $s_1 = 12$.

```
n1.seas <- 12
n2.seas <- 48
trend <- 1:n
id <- 1:n
k.48 <- 1:(n2.seas / 2)
sin.mat <- cos.mat <- matrix(0, nrow = n, ncol = length(k.48))
for (i in 1:length(k.48)) {
  sin.mat[, i] <- sin(2 * pi * id * k.48[i] / n)
  colnames(sin.mat) <- paste("S48", c(1:(length(k.48))), sep = "")
  cos.mat[, i] <- cos(2 * pi * id * k.48[i] / n)
  colnames(cos.mat) <- paste("C48", c(1:(length(k.48))), sep = "")
}
x.mat <- cbind.data.frame(sin.mat, cos.mat, trend)
formula.2.seasons <-
  y ~ 1 +
  sin.mat + cos.mat +
  f(trend, model = "ar1", constr = FALSE)
```

In other situations, when the harmonics corresponding to one seasonal period are not included in the harmonics for another seasonal period, such as for $s_1 = 4$ and $s_2 = 7$, we can include exogenous predictors corresponding to all the distinct harmonics.

2. We can include some (or all) of the suitable harmonics as *dynamic* predictors, and specify the evolution of their coefficients using an rw1 or ar1 specification in the state equation.

```
id.sin <- id.cos <- 1:n
formula.2.seasons.re <-
  y ~ 1 + f(id.sin,sin.mat, model="ar1") +
  f(id.cos,cos.mat, model = "ar1")
```

3. It may be possible to include model="seasonal" once for each seasonal component (code is shown below), although in our experience, the results are not robust:

```
formula <-
  dat ~ 1 +
  f(id1.seas, model = "seasonal", season.length = n.seas.1) +
  f(id2.seas, model = "seasonal", season.length = n.seas.2)
```

5.4 Latent AR(1) model with covariates plus noise

In some situations, we may allow the model for y_t to have a dynamic trend which we wish to model as an ARX(p), i.e., an AR(p) model with exogenous predictors. In practice, what this means is that the response time series y_t is not directly associated with exogenous predictors, but rather, these predictors are associated with the random coefficients that are modeled dynamically, and thus indirectly associated with y_t. In a marketing application for instance, suppose the monthly revenue to a pharmaceutical firm from sales of a drug y_t is modeled via a dynamic trend plus noise, and the dynamic trend is associated with the level of monthly advertising. This is a different model than the one we saw in Section 5.3, where we would have directly studied the association between advertising costs and revenue.

We show the R-INLA feature which allows us to include external predictors in the dynamic AR(1) specification of the state equation as shown below:

$$y_t = x_t + v_t; \quad v_t \sim N(0, \sigma_v^2),$$

$$x_t = \phi x_{t-1} + \sum_{j=1}^{k} \beta_j Z_{t-1,j} + w_t; \quad w_t \sim N(0, \sigma_w^2), \ t = 2, \ldots, n; \qquad (5.6)$$

x_1 follows the normal distribution in (3.3), $|\phi| < 1$, $Z_{t-1,j}$, $j = 1, \ldots, k$ are exogenous predictors, and β_j, $j = 1, \ldots, k$ are the corresponding coefficients. Let the random vector $\boldsymbol{\beta} = (\beta_1, \ldots, \beta_k)'$ have a k-variate normal prior distribution with mean $\mathbf{0}$ and precision matrix $\boldsymbol{Q}$.

For illustrative purposes, we use a custom function (that we sourced at the top of this chapter) to simulate data from an AR(1) process with $k = 2$ predictors, Z_1 and Z_2, each generated from a $N(0,1)$ distribution. We set $\sigma_v^2 = 0.5, \sigma_w^2 = 0.1, \beta_1 = 2, \beta_2 = -0.75$, and $\phi = 0.6$.

```
y.ar1c <-
  simulation.from.ar1c(
    sample.size = 500,
    burn.in = 100,
    phi.ar1 = 0.6,
    ncol.z = 2,
    beta.z = c(2,-0.75),
    V = 0.5,
    W = 0.1,
    plot.data = FALSE,
    seed = 123457
  )
y <- y.ar1c$sim.data$y
Z <- y.ar1c$sim.data$Z
```

We can model this DLM using model= "ar1c" as shown below.

```
f(
  formula,
  model = "ar1c",
  hyper = < hyper > ,
  args.ar1c = list(Z = Z, Q.beta = Q)
)
```

The setup and results for the simulated time series are shown below.

```
idx <- 1:length(y)
Q <- matrix(c(0.1, 0, 0, 0.1), nrow = 2)
```

```
data.ar1c <- cbind.data.frame(y, idx)
formula.ar1c <- y ~ -1 + f(
  idx,
  model = "ar1c",
  args.ar1c = list(Z = Z,
                   Q.beta = Q))
model.ar1c <- inla(
  formula.ar1c,
  data = data.frame(y, idx),
  family = "gaussian",
  control.compute = list(
    dic = TRUE,
    waic = TRUE,
    cpo = TRUE,
    config = TRUE
  )
)
# summary(model.ar1c)
format.inla.out(model.ar1c$summary.hyperpar[,c(1,2)])
##    name                         mean    sd
## 1 Precision for Gaussian obs   1.878  0.165
## 2 Precision for idx            12.605 6.554
## 3 Rho for idx                  0.585  0.009
```

The posterior results show that we recover the true precision parameters and ϕ. Also, as we see below, the posterior means of β_1 and β_2 are close to their true values. Note that since β_1 and β_2 are associated with the random effect x_t, INLA includes their posterior summaries under `summary.random`, after the first n rows corresponding to $x_t, t = 1, \ldots, n$. That is, the latent vector is regarded as $(x_1, x_2, \ldots, x_n, \beta_1, \ldots, \beta_k)$, of length $n + k$.

```
dim(model.ar1c$summary.random$idx)
## [1] 502    8
format.inla.out(model.ar1c$summary.random$idx[501:502, ])
##    ID  mean   sd     0.025q 0.5q   0.975q mode   kld
## 1 501  2.004  0.032  1.942  2.004  2.067  2.004  0
## 2 502 -0.748  0.032 -0.810 -0.748 -0.686 -0.748  0
```

6

Hierarchical Dynamic Models for Panel Time Series

6.1 Introduction

Multilevel or hierarchical models are widely used and have also been implemented in R (Gelman and Hill, 2006). In Chapter 4 of their book, Gómez-Rubio (2020) discuss R-INLA implementation for hierarchical models, including longitudinal data models; also see Krainski et al. (2018). In this chapter, we describe approximate dynamic Bayesian analysis for a set of g univariate time series. Such time series are also referred to as *panel time series* in the literature. Raman et al. (2021b) describe an interesting application in the marketing domain. An alternate hierarchical Bayesian DLM framework for Gaussian time series has been discussed in Gamerman and Migon (1993).

In Section 6.2, we use simulated data to show how to fit models to g time series when we assume that the parameters in the AR(1) state evolution equation are assumed to be the same. Parameters such as the level α in the observation equation can exhibit considerable variation, and may be modeled as a fixed or random effect. Section 6.3 shows a few plausible hierarchical models for NYC ridesourcing usage data.

The custom functions must first be sourced by the user.

```
source("functions_custom.R")
```

We use the following packages.

```
library(tables)
library(tidyverse)
library(INLA)
```

We analyze the following data.

1. Ridesourcing in NYC

6.2 Models with homogenous state evolution

Consider g time series each of length n. Let Y_{it} denote the continuous-valued response at time t for the ith series, $i = 1, \ldots, g$ and $t = 1, \ldots, n$. The response can be written as a data matrix $\boldsymbol{Y}$ with g rows (subjects/locations) and n columns (time points):

$$Y = \begin{pmatrix} y_{11} & y_{12} & \cdots & y_{1n} \\ y_{21} & y_{22} & \cdots & y_{2n} \\ \vdots & \vdots & \ddots & \vdots \\ y_{g1} & y_{g2} & \cdots & y_{gn} \end{pmatrix}. \tag{6.1}$$

In all the models below, v_{it} and w_{it} are uncorrelated, with v_{it} being i.i.d. $N(0, \sigma_v^2)$ and w_{it} being i.i.d. $N(0, \sigma_w^2)$

6.2.1 Example: Simulated homogeneous panel time series with the same level

Consider the following model for y_{it}:

$$y_{it} = \alpha + \beta_{it,1} Z_{it} + v_{it},$$
$$\beta_{it,1} = \phi \beta_{i,t-1,1} + w_{it}, \tag{6.2}$$

where the coefficient $\beta_{it,1}$ corresponding to the exogenous predictor Z_{it} evolves over time as an AR(1) process for each i. The g series have very similar stochastic behavior as indicated by having (nearly) the same values of ϕ, σ_v^2, and σ_w^2. The level of each series is assumed to be the same value α.

```
set.seed(123457)
g <- 100
n <- 250
burn.in <- 500
n.tot <- n + burn.in
alpha <- 10.0
sigma2.w0 <- 0.25
sigma2.v0 <- 1.0
phi.0 <- 0.8
```

The code below generates the response and predictors.

```
yit.all <-
  yit.all.list <-
  z1.all <- z2.all <- z.all <- sigma2.w <- vector("list", g)
for (j in 1:g) {
  yit <- beta.1t   <- z <- vector("double", n.tot)
  u <- runif(1, min = -0.1, max = 0.05)
  phi <- phi.0 + u
  us <- runif(1, min = -0.15, max = 0.25)
  sigma2.w[[j]] <- sigma2.w0 + us
  beta.1t <-
    arima.sim(list(ar = phi, ma = 0), n = n.tot, sd = sqrt(sigma2.w[[j]]))
  z <- rnorm(n.tot)
  v <- rnorm(n.tot)
  for (i in 1:n.tot) {
    yit[i] <- alpha + beta.1t[i]  * z[i] + v[i]
  }
  yit.all.list[[j]] <- tail(yit, n)
  z.all[[j]] <- tail(z, n)
```

```
}
yit.all <- unlist(yit.all.list)
z.all <- unlist(z.all)
```

We set up indexes, replicates, data frame, and the formula.

```
id.beta1 <- rep(1:n, g)
re.beta1  <- rep(1:g, each = n)
data.panelts.1 <-
  cbind.data.frame(id.beta1, z.all, yit.all, re.beta1)
formula.panelts.1 <-
  yit.all ~ 1 +
  f(id.beta1, z.all, model = "ar1", replicate = re.beta1)
```

We fit the model and summarize the results.

```
model.panelts.1 <- inla(
  formula.panelts.1,
  family = "gaussian",
  data = data.panelts.1,
  control.predictor = list(compute = TRUE, link = 1),
  control.compute = list(
    dic = TRUE,
    cpo = TRUE,
    waic = TRUE,
    config = TRUE,
    cpo = TRUE
  )
)
```

```
format.inla.out(model.panelts.1$summary.fixed)
##    name         mean   sd    0.025q 0.5q  0.975q mode   kld
## 1 Intercept  9.999  0.007 9.985  9.999 10.01  9.999  0
format.inla.out(model.panelts.1$summary.hyperpar[,c(1:2)])
##    name                          mean   sd
## 1 Precision for Gaussian obs    1.009  0.012
## 2 Precision for id.beta1        1.302  0.036
## 3 Rho for id.beta1              0.778  0.009
```

The results for the fixed and random parameters show that all the parameters are adequately recovered.

6.2.2 Example: Simulated homogeneous panel time series with different levels

Consider the following model:

$$y_{it} = \alpha_i + \beta_{it,1} Z_{it} + v_{it},$$
$$\beta_{it,1} = \phi \beta_{i,t-1,1} + w_{it}, \tag{6.3}$$

where the coefficient $\beta_{it,1}$ corresponding to the exogenous predictor Z_{it} evolves over time as an AR(1) process for each i. The g series have very similar stochastic behavior as indicated by having (nearly) the same values of ϕ, σ_v^2, and σ_w^2. The levels α_i are assumed to be possibly different across the g series. If g is small, the α_i coefficients may be treated as fixed effects. On the other hand, when g is large, it is usual to assume that α_i are random effects assumed to follow a $N(0, \sigma_\alpha^2)$ distribution, with unknown σ_α^2.

Case 1 Assume that $g = 10$, and α_i are fixed effects. We employ the `fact.alpha <- as.factor(id.alpha)` to set up the g levels to include in the formula. The true parameters `sigma2.w0`, `sigma2.v0`, `phi.0`, `n`, and `burn.in` are the same as in the previous example. The response and predictors are generated as follows.

```
yit.all <-
  yit.all.list <-
  z1.all <- z2.all <- alpha <- z.all <- sigma2.w <- vector("list", g)
for (j in 1:g) {
  yit <- beta.1t  <- z <- vector("double", n.tot)
  u <- runif(1, min = -0.1, max = 0.05)
  phi <- phi.0 + u
  us <- runif(1, min = -0.15, max = 0.25)
  sigma2.w[[j]] <- sigma2.w0 + us
  beta.1t <-
    arima.sim(list(ar = phi, ma = 0), n = n.tot, sd = sqrt(sigma2.w[[j]]))
  alpha[[j]] <- sample(seq(-2, 2, by = 0.1), 1)
  z <- rnorm(n.tot)
  v <- rnorm(n.tot)
  for (i in 1:n.tot) {
    yit[i] <- alpha[[j]] + beta.1t[i]  * z[i] + v[i]
  }
  yit.all.list[[j]] <- tail(yit, n)
  z.all[[j]] <- tail(z, n)
}
yit.all <- unlist(yit.all.list)
z.all <- unlist(z.all)
```

The indexes, replicates, data frame, and the formula follow.

```
id.beta1 <- rep(1:n, g)
re.beta1 <- id.alpha <- rep(1:g, each = n)
fact.alpha <- as.factor(id.alpha)
data.panelts.2 <-
  cbind.data.frame(id.beta1, z.all, yit.all, re.beta1, fact.alpha)
formula.panelts.2 <-
  yit.all ~ -1 + fact.alpha +
  f(id.beta1, z.all, model = "ar1", replicate = re.beta1)
```

We save the output from the fit in `model.panelts.2` using code similar to that in the previous example, but using `formula.panelts.2` and `data.panelts.2` as inputs.

```
model.panelts.2 <- inla(
  formula.panelts.2,
  family = "gaussian",
  data = data.panelts.2,
  control.predictor = list(compute = TRUE, link = 1),
  control.compute = list(
    dic = TRUE,
```

```
      cpo = TRUE,
      waic = TRUE,
      config = TRUE,
      cpo = TRUE
   )
)
```

```
format.inla.out(summary.model.panelts.2$fixed[,c(1,2,4)])
##      name            mean    sd     0.5q
## 1  fact.alpha1    -1.626  0.070  -1.626
## 2  fact.alpha2    -1.569  0.072  -1.569
## 3  fact.alpha3     1.919  0.072   1.919
## 4  fact.alpha4    -0.181  0.070  -0.181
## 5  fact.alpha5    -1.883  0.071  -1.883
## 6  fact.alpha6     1.050  0.070   1.050
## 7  fact.alpha7     0.373  0.070   0.373
## 8  fact.alpha8    -1.023  0.071  -1.023
## 9  fact.alpha9     1.088  0.071   1.088
## 10 fact.alpha10   -0.928  0.072  -0.928
format.inla.out(summary.model.panelts.2$hyperpar[,c(1:2)])
##    name                             mean    sd
## 1 Precision for Gaussian obs       0.997  0.038
## 2 Precision for id.beta1           1.719  0.164
## 3 Rho for id.beta1                 0.765  0.030
```

The posterior means of α_i are close to the true values given by

```
unlist(alpha)
##   [1]   0.4  -0.5   0.2   0.8   1.5  -1.4  -0.6   0.9  -1.1  -0.4
```

Case 2

We next show the code for treating α_i's as independent $N(0, \sigma_\alpha^2)$ random variables, instead of treating them as fixed coefficients. This is useful when g is large; here $g = 100$. In this case, we model the random levels via f(id.alpha, model="iid"), as shown below. The true parameters sigma2.w0, sigma2.v0, phi.0, n, and burn.in are the same as in the previous example. The response and predictors are generated as follows.

```
yit.all <-
  yit.all.list <-
  z1.all <- z2.all <- alpha <- z.all <- sigma2.w <- vector("list", g)
for (j in 1:g) {
  yit <- beta.1t  <- z <- vector("double", n.tot)
  u <- runif(1, min = -0.1, max = 0.05)
  phi <- phi.0 + u
  us <- runif(1, min = -0.15, max = 0.25)
  sigma2.w[[j]] <- sigma2.w0 + us
  beta.1t <-
    arima.sim(list(ar = phi, ma = 0), n = n.tot, sd = sqrt(sigma2.w[[j]]))
  alpha[[j]] <- sample(seq(-2, 2, by = 0.1), 1)
  z <- rnorm(n.tot)
  v <- rnorm(n.tot)
  for (i in 1:n.tot) {
```

```
    yit[i] <- alpha[[j]] + beta.1t[i]  * z[i] + v[i]
  }
  yit.all.list[[j]] <- tail(yit, n)
  z.all[[j]] <- tail(z, n)
}
yit.all <- unlist(yit.all.list)
z.all <- unlist(z.all)
```

The indexes, replicates, data frame, and formula follow.

```
id.beta1 <- rep(1:n, g)
re.beta1 <- id.alpha <- rep(1:g, each = n)
data.panelts.3 <-
  cbind.data.frame(id.beta1, z.all, yit.all, re.beta1, id.alpha)
formula.panelts.3 <-
  yit.all ~ -1 + f(id.alpha, model="iid") +
  f(id.beta1, z.all, model = "ar1", replicate = re.beta1)
```

We save the output from the fit in `model.panelts.3` using code similar to that in the previous example, but using `formula.panelts.3` and `data.panelts.3` as inputs.

```
model.panelts.3 <- inla(
  formula.panelts.3,
  family = "gaussian",
  data = data.panelts.3,
  control.predictor = list(compute = TRUE, link = 1),
  control.compute = list(
    dic = TRUE,
    cpo = TRUE,
    waic = TRUE,
    config = TRUE,
    cpo = TRUE
  )
)
```

```
format.inla.out(model.panelts.3$summary.hyperpar[,c(1:2)])
##    name                            mean  sd
## 1 Precision for Gaussian obs 1.006 0.013
## 2 Precision for id.alpha        0.577 0.075
## 3 Precision for id.beta1        1.349 0.051
## 4 Rho for id.beta1              0.789 0.008
```

Note that in this case, as in the case of treating the levels as fixed effects, we recover the true ϕ, σ_v^2 and σ_w^2. For random α_i, we notice that the estimated posterior precision matches the true precision of the simulated random α's shown below. The overall estimation accuracy should increase as g increases.

```
1/var(unlist(alpha))
## [1] 0.8016
```

In practice, we may often need to model multiple time series for which the stochastic behavior may differ between series. For instance, in (6.3), we may wish to model the state equation as

$$\beta_{it,1} = \phi_i \beta_{i,t-1,1} + w_{it}, \tag{6.4}$$

where w_{it} are $N(0, \sigma^2_{w_i})$ variables. We discuss such models after describing an approach for multivariate time series in Chapter 12.

Remark: More on the `replicate` argument. Recall that we used the `copy` option in Chapter 5. If a model includes several likelihood terms, it may be necessary to share some terms among different parts of the model. In `R-INLA`, we can use the `replicate` feature to enable the linear predictors of the different likelihoods to share the same type of latent effect. Unlike the `copy` feature, `replicate` allows different hyperparameters for each replicate of an effect. The hyperparameter estimation will be based on all the data, but estimates of the latent effects are allowed to vary between the groups.

6.3 Example: Ridesourcing in NYC

Transportation Network Companies (TNCs) provide ridesourcing services by allowing passengers to use their smart phone apps to connect with nearby drivers who typically drive part-time using their own car. We consider weekly ridesourcing usage data from $g = 105$ different taxi zones in NYC from January 2015 until June 2017, providing data for $n = 129$ weeks. The data consists of TNC, Taxi, and Subway usage by taxi zones for each week. TNC usage aggregates three ride modes: Uber, Lyft, and Via. Taxi usage aggregates Yellow and Green Cabs. In most of the zones, TNC usage shows a continuous increase over time, while Taxi usage exhibits a decreasing trend. Subway usage, as expected for NYC, dominates the other two in most zones. The data set also includes data on potential predictors, obtained from various sources such as the *NYC Open Data Portal, National Oceanic and Atmospheric Administration (NOAA)*, the *U.S. Census Bureau*, etc.; see Gerte et al. (2019) and Toman et al. (2020) for details. They include: (a) data that varies by week but remains constant across all taxi zones for any given week, such as holidays (an indicator) and precipitation (inches), and (b) data on socio-economic and land-use variables which we assume vary by taxi zones and consists of Total Population/No. of Buildings, Fulltime Employed/No. of Buildings, Median Age, and Median Earnings. Toman et al. (2020) described time series (ARIMA-GARCH) models to weekly ridesourcing usage within taxi zones in NYC, followed by an investigation into the existence of spatial dependence between the taxi zones. Since the stochastic patterns appear to be varying over time, use of hierarchical dynamic models is a natural choice for such data. Hierarchical models (Gelman and Hill, 2006) enable us to pool information from g groups or individuals (taxi zones in our example) to estimate common coefficients as well as estimating group/zone-specific coefficients. Including time-varying effects in dynamic hierarchical models adds another level of complexity, as we show below.

In the dataset, we have divided each observed count by $k = 10,000$ and split the panel time series data into training (or calibration) and hold-out (or test) portions. Specifically, we set aside $n_h = 5$ weeks from each of the $g = 105$ taxi zones.

```
ride <-
  read.csv("tnc_weekly_data.csv",
           header = TRUE,
           stringsAsFactors = FALSE)
n <- n_distinct(ride$date)
g <- n_distinct(ride$zoneid)
```

```
k <- 10000
ride$tnc <- ride$tnc / k
ride$taxi <- ride$taxi / k
ride$subway <- ride$subway / k
# Train and test data
n.hold <- 5
n.train <- n - n.hold
df.zone <- ride %>%
  group_split(zoneid)
train.df <- df.zone %>%
  lapply(function(x)
    x[1:n.train,]) %>%
  bind_rows()
test.df <- df.zone %>%
  lapply(function(x)
    tail(x, n.hold)) %>%
  bind_rows()
```

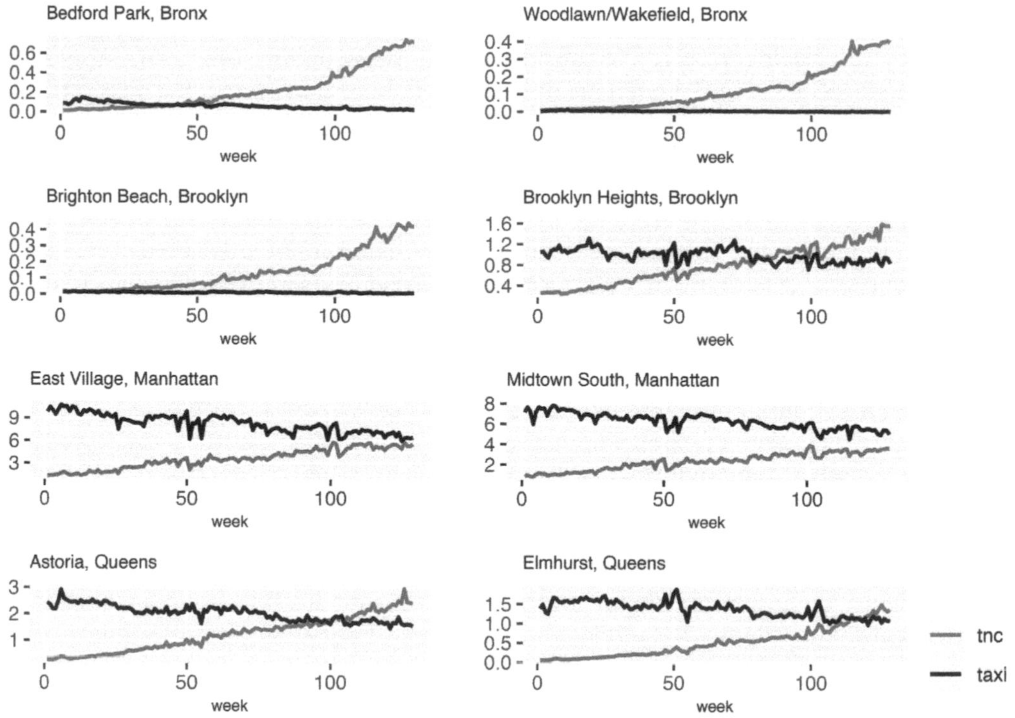

FIGURE 6.1: Scaled TNC and Taxi usage across eight taxi zones.

Figure 6.1 shows the scaled weekly usage of TNC and Taxi in two randomly selected taxi zones from four boroughs in NYC. We see that TNC usage shows an upward trend whereas Taxi usage exhibits an overall downward trend with some fluctuation. There are some similarities and differences in temporal patterns in the different zones, which we explore using hierarchical dynamic models. We first describe the variables used in these models.

Description of variables

The response time series Y_{it} are the scaled TNC usage in week t in zone i, with $n = 129$ and $g = 105$. The response can be written as a data matrix $\boldsymbol{Y}$ with g rows (subjects/locations) and n columns (time points); see (6.1). We consider three types of predictors:

1. Predictors that vary by week t and zone i: these include Subway usage and Taxi usage, denoted by $b_{it,1}$ and $b_{it,2}$; let $\boldsymbol{b}_{it} = (b_{it,1}, b_{it,2})'$. In general, we can assume that there are c such predictors, indexed by $h = 1, \ldots, c$.

2. Predictors that only vary by week t: these variables are constant across all taxi zones i for any given t. They include *holidays* (an indicator) and *precipitation* (inches), which we denote by $d_{t,1}$, and $d_{t,2}$ respectively. Let $\boldsymbol{d}_t = (d_{t,1}, d_{t,2})'$. In general, we can assume that there are a such predictors, indexed by $\ell = 1, \ldots, a$.

3. Predictors that only vary by zone i: these are assumed to stay constant week by week. They include *scaled.population*, *scaled.employment*, *median.age*, and *median.earnings*, and are denoted by $s_{i,1}, s_{i,2}, s_{i,3}$, and $s_{i,4}$ respectively; let $\boldsymbol{s}_i = (s_{i,1}, s_{i,2}, s_{i,3}, s_{i,4})'$. In general, we can assume that there are b such predictors.

Below, we describe and fit the most general hierarchical dynamic model for the ridesourcing data, i.e., Model H1.

6.3.1 Model H1. Dynamic intercept and exogenous predictors

Consider the following model for Y_{it}:

$$
\begin{aligned}
Y_{it} &= \alpha + \beta_{it,0} + \boldsymbol{b}'_{it}\boldsymbol{\beta}_{it} + \boldsymbol{d}'_t\boldsymbol{\gamma}_t + \boldsymbol{s}'_i\boldsymbol{\eta}_i + v_{it} \\
\beta_{it,0} &= \phi_0\beta_{i,t-1,0} + w_{it,0}, \\
\beta_{it,h} &= \phi_h^{(\beta)}\beta_{i,t-1,h} + w_{it,h}^{(\beta)}, \ h = 1, 2, \\
\gamma_{t,\ell} &= \phi_\ell^{(\gamma)}\gamma_{t-1,\ell} + w_{t,\ell}^{(\gamma)}, \ \ell = 1, 2.
\end{aligned}
\tag{6.5}
$$

We have included a level α, and a time-varying intercept $\beta_{it,0}$ which is modeled as a latent zero-mean AR(1) process with coefficient ϕ_0. The coefficients corresponding to Subway usage and Taxi usage are $\beta_{it,h}$, $h = 1, 2$, each of which evolves as a latent Gaussian AR(1) process with coefficients $\phi_1^{(\beta)}$ and $\phi_2^{(\beta)}$ respectively. The coefficients corresponding to holidays and precipitation are $\gamma_{t,\ell}$, $\ell = 1, 2$, and evolve as latent Gaussian AR(1) processes with coefficients $\phi_1^{(\gamma)}$ and $\phi_2^{(\gamma)}$ respectively. The $\boldsymbol{\eta}_i$ coefficients corresponding to the demographic and land-use predictors are only taxi-zone specific and do not follow any dynamic model. We assume that all the AR(1) coefficients lie between -1 and 1. The observation errors v_{it} are assumed to be N$(0, \sigma_v^2)$. The state errors $w_{it,0}$ are N$(0, \sigma_{w,0}^2)$, the errors $w_{it,h}^{(\beta)}$ are N$(0, \sigma_{w,\beta_h}^2)$, $h = 1, 2$, and the errors $w_{t,\ell}^{(\gamma)}$ are N$(0, \sigma_{w,\gamma_\ell}^2)$, $\ell = 1, 2$. The errors are uncorrelated. For elements of $\boldsymbol{\eta}_i$ in the model, we assume independent normal priors. An important aspect of fitting Model H1 is the creation of indexes to handle the different types of coefficients described above (also see Chapter 6 of Gómez-Rubio (2020)).

Model H1: indexes and replicates

1. **Indexes for time and zone varying coefficients**. The time index *id.b0* relates to the intercept $\beta_{it,0}$ which varies by week t and zone i. To capture the time-varying effect, we use `rep(1:n,g)`, which repeats the weeks $1, \ldots, n$ for all $g = 105$ zones.

To allow the intercept to vary by zones, we use the `replicate` argument in the function `f()`. Since the first set of times $1, \ldots, n$ belongs to the first zone, $i = 1$, the index has 1 repeated n times; the next set of $1, \ldots, n$ is for zone $i = 2$, so that the index is 2 repeated n times, and so on. This is represented by `rep(1:g, each = n)` for *re.b0* in the formula. We similarly set up the indexes *id.subway* and *id.taxi*, as well as the replicates *re.subway* and *re.taxi* corresponding to $\beta_{it,1}$ and $\beta_{it,2}$ respectively.

2. **Indexes for time-varying coefficients.** For the predictors *holidays* and *precip*, the γ coefficients vary only by time t and not by zone i. The indexes *id.holiday* and *id.precip* are defined similar to *id.b0*, i.e., by `rep(1:n,g)`. However, the `replicate` argument must be different. We define *re.holiday* as a column of 1's repeated $n \times g$ times, i.e., by `rep(1, (n*g))`. In addition, we allow a large, fixed precision for the hyperparameter. The index and replicate arguments for *precip* are set up similarly.

3. **Indexes for zone specific coefficients.** Since the η_i coefficients corresponding to the demographic and land use predictors do not vary over time, but are zone specific, we do not need to set up indexes similar to *id.b0*. However, to indicate that they are random effects (by zone), we need to set up the replicate arguments *re.age*, *re.earn*, *re.pop*, and *re.emp* as repeating the sequence $1, \ldots, g$ for each t, i.e., by `,rep(1:g, each = n)`. These indexes and replicates for Model H1 are shown in the code below.

```
id.b0 <- id.subway <- id.taxi <- rep(1:n, g)
id.holiday <- id.precip <- rep(1:n, g)
re.b0 <- re.subway <- re.taxi <- rep(1:g, each = n)
re.holiday <- re.precip <- rep(1, (n * g))
re.age <- re.earn <- re.pop <- re.emp <- rep(1:g, each = n)
```

We merge these with the scaled data to create a data frame `ride.data` that is used in the `inla()` fit.

```
ride.data <-
  cbind.data.frame(
    ride,id.b0,id.subway,id.taxi,
    id.holiday,id.precip,re.b0,
    re.subway,re.taxi,re.holiday,
    re.precip,re.age,re.earn,re.pop,re.emp
  )
```

Model H1: formula and model fitting

```
formula.tnc.H1 <- tnc ~ 1 +
  f(id.b0, model = "ar1", replicate = re.b0) +
  f(id.subway, subway, model = "ar1", replicate = re.subway) +
  f(id.taxi, taxi, model = "ar1", replicate = re.taxi) +
  f(
    id.holiday,
    holidays,
    model = "ar1",
    replicate = re.holiday,
```

```
  hyper = list(theta = list(initial = 20, fixed = TRUE))
) +
f(
  id.precip,
  precip,
  model = "ar1",
  replicate = re.precip,
  hyper = list(theta = list(initial = 20, fixed = TRUE))
) +
f(re.age, median.age, model = "iid") +
f(re.earn, median.earnings, model = "iid") +
f(re.pop, scaled.population, model = "iid") +
f(re.emp, scaled.employment, model = "iid")
```

We have used default R-INLA priors for all parameters. The condensed output shown below consists of posterior summaries from Model H1.

```
model.tnc.H1 <- inla(
  formula.tnc.H1,
  data = ride.data,
  control.predictor = list(compute = TRUE),
  control.compute = list(
    dic = TRUE,
    waic = TRUE,
    cpo = TRUE,
    config = TRUE
  )
)
format.inla.out(summary.model.tnc.H1$fixed)
format.inla.out(model.tnc.H1$summary.hyperpar[,c(1:2)])


##   name        mean sd    0.025q 0.5q 0.975q mode kld
## 1 Intercept -0.4 0.146 -0.686 -0.4 -0.114 -0.4 0

##    name                          mean      sd
## 1  Precision for Gaussian obs   1.995e+03 7.805e+01
## 2  Precision for id.b0          2.156e+00 7.540e-01
## 3  Rho for id.b0                9.990e-01 0.000e+00
## 4  Precision for id.subway      6.815e+03 1.298e+03
## 5  Rho for id.subway            9.980e-01 0.000e+00
## 6  Precision for id.taxi        1.690e+00 3.520e-01
## 7  Rho for id.taxi              9.990e-01 0.000e+00
## 8  Precision for id.holiday     6.484e+02 3.195e+02
## 9  Rho for id.holiday           3.280e-01 4.610e-01
## 10 Precision for id.precip      2.420e+02 6.784e+01
## 11 Rho for id.precip            2.470e-01 1.030e-01
## 12 Precision for re.age         9.849e+03 5.734e+03
## 13 Precision for re.earn        1.068e+06 1.660e+05
## 14 Precision for re.pop         3.820e+03 2.679e+03
## 15 Precision for re.emp         4.898e+05 8.218e+04
```

From the output, we see that Model H1 estimates a significant negative level of -0.4. The AR(1) coefficient estimates associated with the β coefficients for Taxi and Subway usage and the intercept β_0 are concentrated in the vicinity of 0.99, (suggesting perhaps one can use a random walk model for these parameters). Furthermore, the estimated posterior precisions are high for most of the coefficients, suggesting that these can be treated as fixed effects (static), leading to simpler models than Model H1. We explore these scenarios below.

Note that if the level α had been insignificant, we could fit a model excluding the level, by setting `formula.tnc.H1 <- tnc ~ -1 + ...` in the formula, keeping all other entries the same as above. We do not show these results.

For the model with level α, Figure 6.2 compares observed and fitted values for the eight different zones that we discussed earlier. This plot indicates possible overfitting in all zones, as expected from the high `Precision for the Gaussian observations` in the output. Recall that the `config = TRUE` option enables us to obtain samples from approximate posterior distributions (see Section 3.10). Using the default argument `line.type = "solid"` in the function `multiline.plot()` provides us with an overlay plot of the actual and fitted values. In order to distinguish them, we have used `line.type = "dotdash"`.

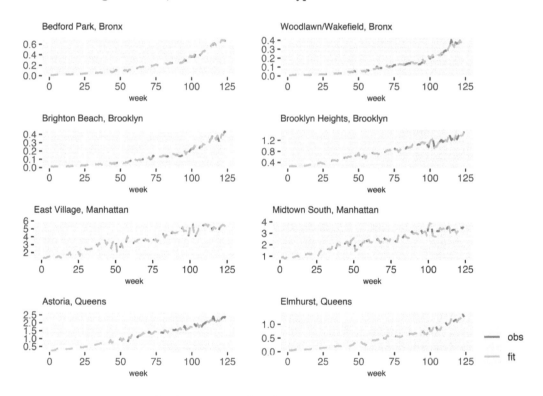

FIGURE 6.2: Observed (red) and fitted (blue) values from Model H1 for eight taxi zones.

In the rest of the chapter, we consider simpler models for TNC usage across zones and over time.

6.3.2   Model H2. Dynamic intercept and Taxi usage

In Model H2, we consider Subway usage and other exogenous variables as fixed effects, setting only the intercept and Taxi usage to be random, dynamic effects. Similar to Model H1, the estimated AR(1) coefficients in the state equations for the intercept and Taxi usage were close to 1, so that we show random walk evolutions for these state variables below. The overall level α is assumed to be a fixed value which is common across the g zones.

$$Y_{it} = \alpha + \beta_{it,0} + \beta_{it,1}b_{it,1} + \beta_2 b_{it,2} + \boldsymbol{d}_t'\boldsymbol{\gamma} + \boldsymbol{s}_i'\boldsymbol{\eta} + v_{it},$$
$$\beta_{it,0} = \beta_{i,t-1,0} + w_{it,0},$$
$$\beta_{it,1} = \beta_{i,t-1,1} + w_{it,1}^{(\beta)}. \tag{6.6}$$

The formula and model fit are set up as follows.

```
formula.tnc.H2 <- tnc ~ 1 + subway + holidays + precip +
  median.age + median.earnings + scaled.population + scaled.employment +
  f(id.b0,
    model = "rw1",
    constr = FALSE,
    replicate = re.b0) +
  f(id.taxi,
    taxi,
    model = "rw1",
    constr = FALSE,
    replicate = re.taxi)
model.tnc.H2 <- inla(
  formula.tnc.H2,
  data = ride.data,
  control.predictor = list(compute = TRUE),
  control.compute = list(
    dic = TRUE,
    waic = TRUE,
    cpo = TRUE,
    config = TRUE
  )
)
format.inla.out(model.tnc.H2$summary.fixed[,c(1:5)])
format.inla.out(model.tnc.H2$summary.hyperpar[,c(1:2)])
```

```
##    name                    mean    sd       0.025q    0.5q    0.975q
## 1 Intercept              -2.736 559.697 -1102.662 -2.741 1096.100
## 2 subway                  0.002   0.000     0.002  0.002    0.003
## 3 holidays                0.007   0.001     0.005  0.007    0.009
## 4 precip                  0.004   0.000     0.003  0.004    0.005
## 5 median.age             -0.004   0.005    -0.013 -0.004    0.006
## 6 median.earnings         0.000   0.000     0.000  0.000    0.000
## 7 scaled.population      -0.004   0.006    -0.016 -0.004    0.009
## 8 scaled.employment       0.000   0.000     0.000  0.000    0.000
##    name                         mean sd
## 1 Precision for Gaussian obs    1411 52.21
## 2 Precision for id.b0           1096 37.43
## 3 Precision for id.taxi         1252 45.33
```

Among the fixed effects, only Subway usage, holidays, precipitation, and scaled employment (see the result without `format.inla.out`) are significant. While we only show a few significant digits here, a more complete output can be obtained via `summary(model.tnc.H2)`. Figure 6.3 shows the observed and fitted usage from Model H2 for eight taxi zones.

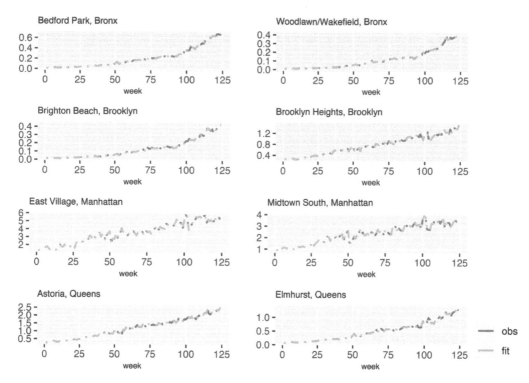

FIGURE 6.3: Observed (red) and fitted (blue) values from Model H2 for eight taxi zones.

In many situations, it may be naive to assume that the level is the same for all zones. To address this, we let the level vary across zones in Model H2, so that the observation equation in **Model H2b** becomes

$$Y_{it} = \alpha_0 + \alpha_i + \beta_{it,0} + \beta_{it,1}b_{it,1} + \beta_2 b_{it,2} + \boldsymbol{d}_t'\boldsymbol{\gamma} + \boldsymbol{s}_i'\boldsymbol{\eta} + v_{it}, \qquad (6.7)$$

while the state equations are the same as under Model H2. We assume that $\alpha_i \sim N(0, \sigma_\alpha^2)$, a random effect across the g zones. Including +1 in the formula below allows us to include an overall fixed level α_0.

```
formula.tnc.H2b <- tnc ~ 1 + subway + holidays + precip +
  median.age + median.earnings + scaled.population + scaled.employment +
  f(id.alpha, model="iid") +
  f(id.taxi,
    taxi,
    model = "rw1",
    constr = FALSE,
    replicate = re.taxi)
```

We show a summary of results below. The precision for α_i is significantly higher than zero, implying that there may be some variation in the level between zones.

```
format.inla.out(model.tnc.H2b$summary.fixed[,c(1:5)])
format.inla.out(model.tnc.H2b$summary.hyperpar[,c(1,2)])
```

```
##    name                  mean    sd    0.025q  0.5q   0.975q
## 1 Intercept            -3.422 0.214  -3.846 -3.421 -3.005
## 2 subway                0.006 0.001   0.004  0.006  0.007
## 3 holidays             -0.007 0.003  -0.012 -0.007 -0.002
## 4 precip                0.015 0.001   0.012  0.015  0.017
## 5 median.age            0.072 0.003   0.065  0.072  0.078
## 6 median.earnings       0.000 0.000   0.000  0.000  0.000
## 7 scaled.population    -0.021 0.004  -0.030 -0.021 -0.012
## 8 scaled.employment     0.001 0.000   0.001  0.001  0.001
##    name                            mean      sd
## 1 Precision for Gaussian obs   108.368    2.011
## 2 Precision for id.alpha         0.572    0.088
## 3 Precision for id.taxi        531.451   24.098
```

Figure 6.4 shows observed and fitted values from Model H2b.

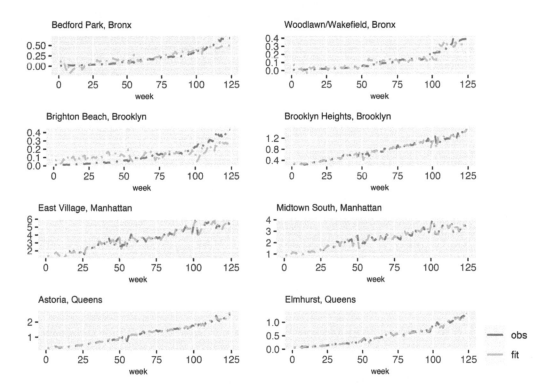

FIGURE 6.4: Observed (red) and fitted (blue) values from Model H2b for eight taxi zones.

Model comparisons are shown in Table 6.1. The data appears to prefer Model H2 as compared with Model H2b. Therefore, we will assume a fixed α across g zones in the following models.

Remark. If the number of zones g was small, we can model the α_i as fixed effects as shown in the model formula for simulated data in Section 6.2.2, using `zone.alpha = as.factor(id.alpha)`.

6.3.3   Model H3. Taxi usage varies by time and zone

In this model, the intercept is assumed to be a fixed effect and is merged into the level α. Subway usage, median age, median earnings, scaled population, and scaled employment are other fixed effects. Taxi usage is assumed to vary over time and by zone, while holiday, and precipitation are assumed to only vary over time; these effects are modeled by random walks. While we do not show the code and output (see the GitHub page), Figure 6.5 shows the observed and fitted TNC usage from Model H3 for eight taxi zones.

$$Y_{it} = \alpha + \beta_{it,1}b_{it,1} + \beta_2 b_{it,2} + \boldsymbol{d}'_t\boldsymbol{\gamma}_t + \boldsymbol{s}'_i\boldsymbol{\eta}_i + v_{it},$$
$$\beta_{it,1} = \beta_{i,t-1,1} + w^{(\beta)}_{it,1},$$
$$\gamma_{t,\ell} = \gamma_{t-1,\ell} + w^{(\gamma)}_{t,\ell}, \ \ell = 1, 2. \tag{6.8}$$

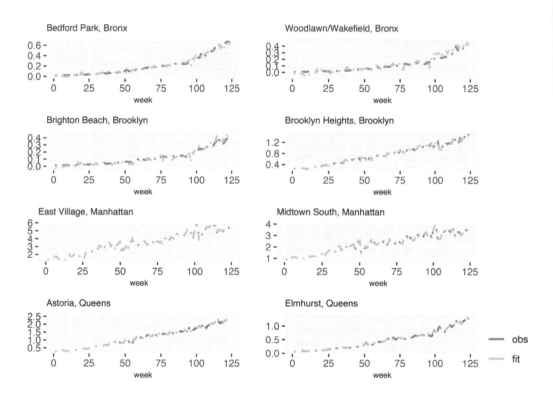

FIGURE 6.5: Observed (red) and fitted (blue) values from Model H3 for eight taxi zones.

6.3.4   Model H4. Fixed intercept, Taxi usage varies over time and zones

Since the results from Model H1 shows high precisions for state errors corresponding to holiday and precipitation, we now treat these variables as static, fixed effects. That is, only Taxi usage varies over time and zones.

$$Y_{it} = \alpha + \beta_{it,1}b_{it,1} + \beta_2 b_{it,2} + \boldsymbol{d}'_t\boldsymbol{\gamma} + \boldsymbol{s}'_i\boldsymbol{\eta} + v_{it},$$
$$\beta_{it,1} = \beta_{i,t-1,1} + w^{(\beta)}_{it,1}. \tag{6.9}$$

Figure 6.6 shows the observed and fitted usage from Model H4 for the eight taxi zones.

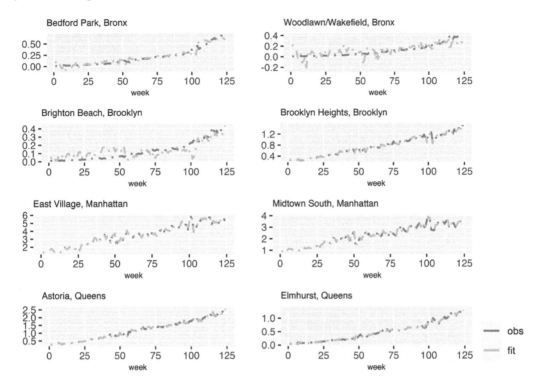

FIGURE 6.6: Observed (red) and fitted (blue) values from Model H4 for eight taxi zones.

6.4 Model comparison

We compare the five models with respect to their in-sample and out of sample performance. In so doing, we first show plots of observed versus fitted values obtained from the five models for a specific zone (Brighton Beach) in Figure 6.7. We also compute and display the in-sample MAPE based on training data for each model for Brighton Beach in Figure 6.7. As suggested from the plot, the MAPE value from Model H1 is the lowest.

In-sample model comparisons via DIC, WAIC, and PsBF, along with MAE and MAPE across zones are shown in Table 6.1. Note that the MAE (or MAPE) in the table is the average of the MAE (or MAPE) values across zones. The out-of-sample model comparisons via MAE and MAPE (computed based on hold-out data) are shown in Table 6.2. We can also look at the distribution of MAPE and MAE values from all zones and compare models. Figure 6.8 shows boxplots of out-of-sample MAPE values for all zones from the five models; we have suppressed outliers in the boxplots for visual clarity. To see the outliers, set `outline = TRUE`. The figure shows that the median of MAPE values across zones is lowest from Model H2.

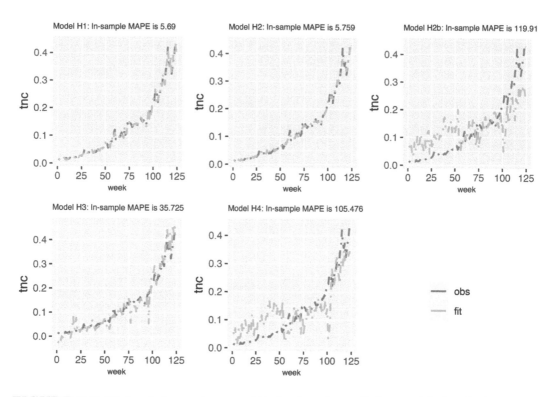

FIGURE 6.7: Plots of observed versus fitted values from all five models for the Brighton Beach taxi zone.

TABLE 6.1: In-sample model comparisons.

Model	DIC	WAIC	PsBF	Average in-sample MAE	Average in-sample MAPE
Model H1	−56232.03	−56568.628	106.996	0.007	2.798
Model H2	−51565.851	−51916.779	96.177	0.008	2.937
Model H2b	−21612.062	−22698.194	70.678	0.046	40.531
Model H3	−33653.708	−35396.272	12.06	0.02	20.346
Model H4	−21371.145	−22998.43	7.905	0.038	46.034

TABLE 6.2: Out-of-sample model comparisons.

Model	Average out-of-sample MAE	Average out-of-sample MAPE
Model H1	0.021	2.635
Model H2	0.022	1.861
Model H2b	0.112	13.470
Model H3	0.060	11.318
Model H4	0.102	12.666

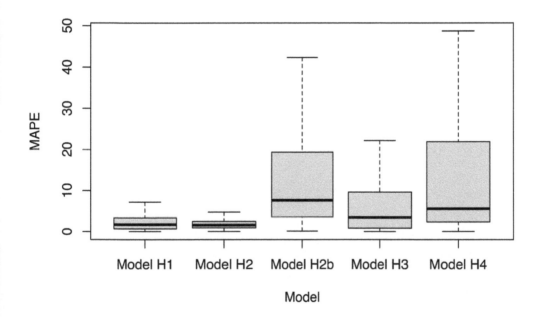

FIGURE 6.8: Boxplots of out-of-sample MAPE values across zones from five models.

7

Non-Gaussian Continuous Responses

7.1   Introduction

In this chapter, we describe state space models for non-Gaussian time series. We describe a gamma state space model in Section 7.2, and consider a Weibull state space model in Section 7.3. In Section 7.4, we present a beta state space model for time series of proportions.

The custom functions must first be sourced by the user.

```
source("functions_custom.R")
```

The list of packages used in this chapter are as follows.

```
library(INLA)
library(astsa)
library(readxl)
library(lubridate)
library(tidyverse)
```

We analyze the following data

1. Volatility index (VIX) times series

2. Crest market share time series.

7.2   Gamma state space model

Let $Y_t \sim \text{Gamma}(\tau, \tau/\mu_t)$ where $\tau > 0$, and $\mu_t > 0$. This parametrization implies that

$$E[Y_t|\tau, \mu_t] = \mu_t,$$

and

$$\text{Var}[Y_t|\tau, \mu_t] = \mu_t^2/\tau.$$

Thus, the precision parameter $\tau > 0$ is static, whereas the mean μ_t is time-varying. We assume that

$$\eta_t = \log(\mu_t) = \alpha + \boldsymbol{f}_t' \boldsymbol{x}_t \tag{7.1}$$

and

$$x_t = Gx_{t-1} + w_t, \tag{7.2}$$

where α is an intercept, x_t is an h-dimensional state vector, f_t is an h-dimensional covariate vector, G is an $h \times h$ transition matrix, and $w_t \sim N(0, W)$. Note that if we do not have any exogenous predictors, then $\eta_t = \log(\mu_t) = \alpha + x_t$, where x_t is a scalar and the state equation will reduce to

$$x_t = G x_{t-1} + w_t. \tag{7.3}$$

7.2.1 Example: Volatility index (VIX) time series

We use the monthly VIX data for a period of 5 years from October 2012 to October 2017 (Aktekin et al., 2020); as noted in this paper, "VIX is the ticker symbol for the Chicago Board Options Exchange volatility index, which shows the market's expectation of a 30-day volatility. It is constructed using the implied volatilities of a wide range of S&P500 index options." Figure 7.1 shows a time series plot of the VIX values.

```
index <- read.csv("VIX.csv", header = TRUE)
vix <- index[, 1]
n <- nrow(index)
tsline(
  as.numeric(vix),
  ylab = "VIX",
  xlab = "t",
  line.color = "red",
  line.size = 0.6
) +
  ylim(9, 30) +
  scale_x_continuous(
    breaks = c(1, 13, 25, 37, 49, 60),
    labels = c("2012", "2013", "2014", "2015", "2016", "2017")
  )
```

Model G1

We first consider a model without regressors to explain the VIX time series. Specifically, assuming that f_t' is 1, so that x_t denotes a random state, we write this model as

$$
\begin{aligned}
Y_t | \mu_t, \tau &\sim \mathrm{Gamma}(\tau, \tau/\mu_t), \\
\eta_t &= \log(\mu_t) = \alpha + x_t, \\
x_t &= \phi x_{t-1} + w_t.
\end{aligned} \tag{7.4}
$$

The formula and code for fitting the model are shown below. Figure 7.2 shows the observed and fitted VIX values from the fitted Model G1.

```
id.x <- 1:length(vix)
vix.data <- cbind.data.frame(vix, id.x)
formula.vix.G1 <-  vix ~  1 + f(id.x, model = "ar1")
model.vix.G1 <-  inla(
  formula.vix.G1,
```

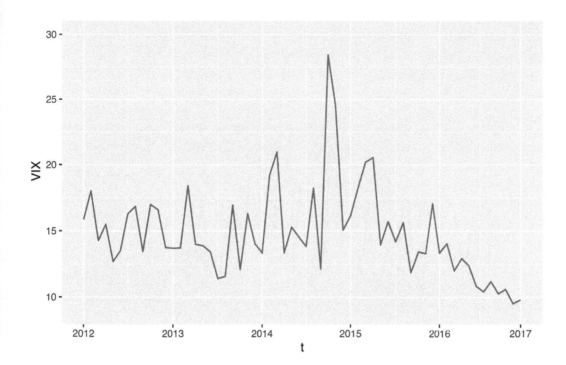

FIGURE 7.1: Monthly time series of VIX. The x-axis labels show the years for the data series.

```
  data = vix.data,
  family = "gamma",
  control.predictor = list(compute = TRUE),
  control.compute = list(
    dic = TRUE,
    waic = TRUE,
    cpo = TRUE,
    config = TRUE
  )
)
# summary(model.vix.G1)
format.inla.out(model.vix.G1$summary.fixed[,c(1:5)])
##    name        mean  sd    0.025q 0.5q 0.975q
## 1 Intercept 2.654 0.102 2.418  2.66 2.845
format.inla.out(model.vix.G1$summary.hyperpar[,(1:2)])
##    name                              mean    sd
## 1 Precision parameter for Gamma obs 47.137 11.941
## 2 Precision for id.x                42.556 24.436
## 3 Rho for id.x                      0.897  0.073
```

Note that the summary for the precision parameter τ is given by "Precision parameter for Gamma obs" in `model.vix.G1$summary.hyperpar`. We can also obtain samples from the posterior distribution of τ using `inla.hyperpar.sample()`.

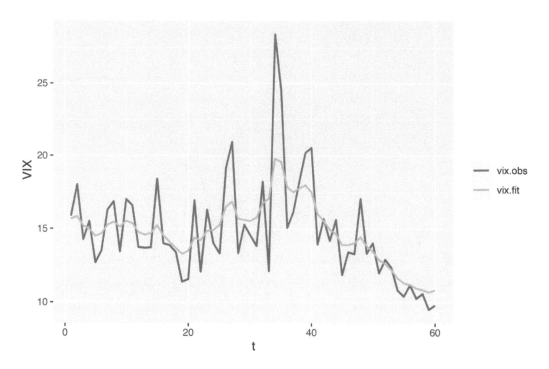

FIGURE 7.2: Observed (red) and fitted (blue) VIX values from the model without regressors (Model G1).

Model G2

As an alternative model, we can consider including another volatility index, VXN, as a regressor in the model. As noted by Aktekin et al. (2020), "VXN is a measure of market expectations of 30-day volatility for the Nasdaq-100 market index, as implied by the price of near-term options on this index." Figure 7.3 shows the overlay plot of the time series VIX and VXN, which suggests that the two indexes are positively correlated.

```
vxn <- index[, 2]
plot.df <- as_tibble(cbind.data.frame(
  time = id.x,
  vix.obs = vix.data$vix,
  vxn = vxn
))
multiline.plot(
  plot.df,
  title = "",
  xlab = "t",
  ylab = "VIX",
  line.size = 0.8,
  line.color = c("red", "black")
)
```

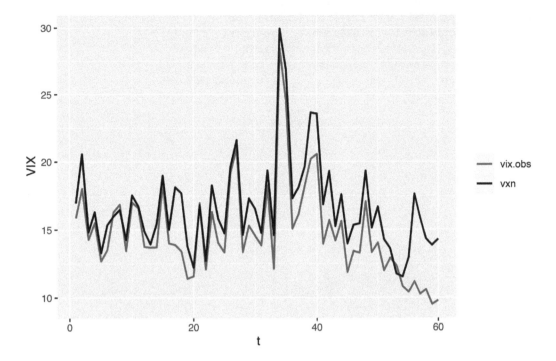

FIGURE 7.3: Monthly time series of VIX (red) and VXN (black).

Model G2 can be represented as

$$
\begin{aligned}
Y_t|\mu_t, \tau &\sim \mathrm{Gamma}(\tau, \tau/\mu_t), \\
\eta_t &= \alpha + \beta_t Z_t, \\
\beta_t &= \phi\beta_{t-1} + w_t.
\end{aligned}
\tag{7.5}
$$

Here, $\boldsymbol{f}_t$ is the VXN variable Z_t. The results from this model are shown below, while observed and fitted VIX values are shown in Figure 7.4.

```
id.vxn <- id.x
formula.vix.G2 <- vix ~ 1 + f(id.vxn, vxn, model = "ar1")
model.vix.G2 <- inla(
  formula.vix.G2,
  data = vix.data,
  family = "gamma",
  control.predictor = list(compute = TRUE),
  control.compute = list(
    dic = TRUE,
    waic = TRUE,
    cpo = TRUE,
    config = TRUE
  )
)
# summary(model.vix.G2)
format.inla.out(model.vix.G2$summary.fixed[, c(1:5)])
```

```
##   name       mean  sd     0.025q 0.5q  0.975q
## 1 Intercept 1.835 0.061 1.715  1.834 1.956
format.inla.out(model.vix.G2$summary.hyperpar[, c(1:2)])
##   name                                mean    sd
## 1 Precision parameter for Gamma obs  373.738 112.922
## 2 Precision for id.vxn              1651.985 932.790
## 3 Rho for id.vxn                       0.994   0.004
```

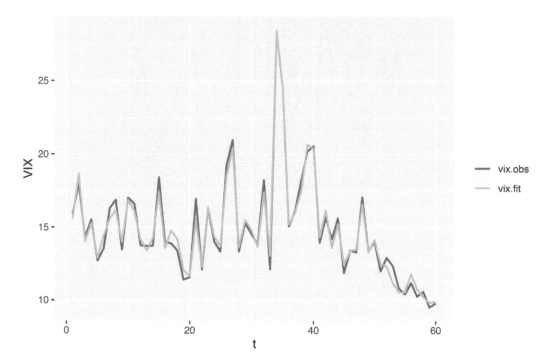

FIGURE 7.4: Observed (red) and fitted (blue) VIX values from model with VXN as a dynamic regressor (Model G2).

Model G3

As a third example, we fit a model with a dynamic intercept term and a fixed effect corresponding to the regressor VXN:

$$
\begin{aligned}
Y_t|\mu_t, \tau &\sim \text{Gamma}(\tau, \tau/\mu_t), \\
\eta_t &= \alpha + \beta Z_t + x_t, \\
x_t &= \phi x_{t-1} + w_t;
\end{aligned}
\tag{7.6}
$$

here, $\boldsymbol{f}_t$ is 1. The code and the results from this model fit are given below.

```
formula.vix.G3 <-  vix ~ 1 + vxn + f(id.x, model = "ar1")
model.vix.G3 <- inla(
  formula.vix.G3,
  data = vix.data,
  family = "gamma",
```

```
  control.predictor = list(compute = TRUE),
  control.compute = list(
    dic = TRUE,
    waic = TRUE,
    cpo = TRUE,
    config = TRUE
  )
)
# summary(model.vix.G3)
format.inla.out(model.vix.G3$summary.fixed[, c(1:5)])
##   name        mean   sd    0.025q  0.5q   0.975q
## 1 Intercept  1.767  0.079  1.606   1.768  1.923
## 2 vxn        0.053  0.003  0.047   0.053  0.060
format.inla.out(model.vix.G3$summary.hyperpar[, c(1:2)])
##   name                                   mean     sd
## 1 Precision parameter for Gamma obs  390.501  133.029
## 2 Precision for id.x                  97.510   41.649
## 3 Rho for id.x                         0.882    0.068
```

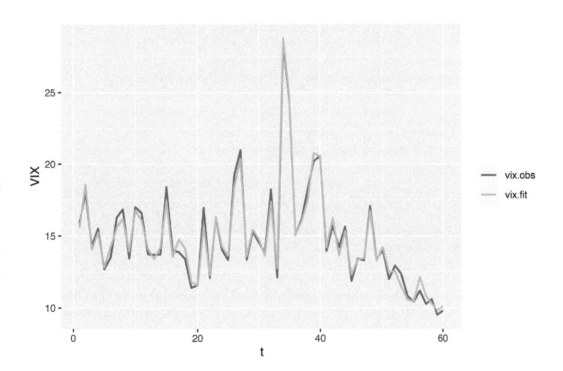

FIGURE 7.5: Observed (red) and fitted (blue) VIX values from a dynamic intercept and fixed VXN (Model G3).

In Table 7.1, we present a comparison of the three models. Based on the marginal likelihood as well as the DIC and WAIC, we see that Model G2 with a dynamic covariate effect does slightly better in terms of fit than Model G3, which has a dynamic intercept term but a static covariate effect.

TABLE 7.1: Comparison of Models G1, G2, and G3.

Model	DIC	WAIC	PsBF	Marginal likelihood
Model G1	277.208	280.61	-142.295	-155.076
Model G2	156.997	155.313	-81.417	-104.236
Model G3	157.677	154.716	-82.49	-112.027

As an alternative observation model for the VIX volatility index, we can consider a Weibull model, as discussed in the next section.

7.3 Weibull state space model

We can define a state space model for Weibull distributed observations. Let $Y_t \sim$ Weibull(γ_w, λ_t), where $\gamma_w > 0$ is the static Weibull shape parameter and $\lambda_t > 0$ is the dynamic scale parameter. R-INLA uses two different parameterizations for the Weibull density. The parameterization referred to as "variant 0" uses the following form of the Weibull density

$$p(y_t|\gamma_w, \lambda_t) = \gamma_w \lambda_t y_t^{\gamma_w - 1} e^{-\lambda_t y_t^{\gamma_w}}.$$

The alternative parameterization referred to as "variant 1" uses the density

$$p(y_t|\gamma_w, \lambda_t) = \gamma_w \lambda_t^{\gamma_w} y_t^{\gamma_w - 1} e^{-(\lambda_t y_t)^{\gamma_w}}.$$

For both parameterizations, the linear predictor is defined as

$$\eta_t = \log(\lambda_t) = \alpha + \boldsymbol{f}_t' \boldsymbol{x}_t, \text{ with,}$$
$$\boldsymbol{x}_t = \boldsymbol{G} \boldsymbol{x}_{t-1} + \boldsymbol{w}_t, \tag{7.7}$$

where $\boldsymbol{x}_t$ is an h-dimensional state vector, $\boldsymbol{f}_t$ is a $h \times 1$ covariate vector, $\boldsymbol{G}$ is an $h \times h$ transition matrix, and $\boldsymbol{w}_t \sim N(\boldsymbol{0}, \boldsymbol{W})$.

Model W1

We first fit a model with a dynamic intercept, but without any regressors. The model has the same structure as Model G1. The observed and fitted VIX values are shown in Figure 7.6. The choice of model variant is given in `control.family`.

```
formula.vix.W1 <- vix ~ 1 + f(id.x, model = "ar1")
model.vix.W1 <- inla(
  formula.vix.W1,
  data = vix.data,
  family = "weibull",
  control.family = list(variant = 0),
  control.predictor = list(compute = TRUE, link = 1),
  control.compute = list(
    dic = TRUE,
    waic = TRUE,
```

```
    cpo = TRUE,
    config = TRUE
  )
)
# summary(model.vix.W1)
format.inla.out(model.vix.W1$summary.fixed[, c(1:5)])
##    name         mean    sd    0.025q  0.5q    0.975q
## 1 Intercept  -22.65  2.723  -30.6  -21.88  -19.2
format.inla.out(model.vix.W1$summary.hyperpar[, c(1:2)])
##    name                           mean   sd
## 1 alpha parameter for weibull   9.182  1.039
## 2 Precision for id.x            0.282  0.108
## 3 Rho for id.x                  0.849  0.088
```

Note that γ_w in our notation corresponds to "alpha parameter for Weibull" in the `inla()` output. For obtaining fitted values, the standard procedure of using `model.vix.W1$summary.fitted.values$mean` will lead to incorrect fits. The suggested procedure to obtain fits for the Weibull family is to use `inla.posterior.sample()` as shown below. Discussion on the function `fun.pred()` follows after the code.

```
post.sample.W1 <-
  inla.posterior.sample(n = 1000, model.vix.W1, seed = 123457)
fun.pred <- function(x) {
  return(gamma(1 + 1 / x$hyperpar[1]) /
         (exp(x$latent[1:n]) ^ (1 / x$hyperpar[1])))
}
fits <- post.sample.W1 %>%
  sapply(function(x)
    fun.pred(x))
fit.vix.W1 <- rowMeans(fits)
plot.df <- as_tibble(cbind.data.frame(
  time = id.x,
  vix.obs = vix.data$vix,
  vix.fit = fit.vix.W1
))
multiline.plot(
  plot.df,
  title = "",
  xlab = "t",
  ylab = "VIX",
  line.size = 0.8,
  line.color = c("red", "blue")
)
```

Since we have used the parameterization `variant=0` in Model W1, the fitted values are obtained using the transformation

$$E(y_t|\gamma_w, \lambda_t) = \lambda_t^{-1/\gamma_w} \Gamma(1 + 1/\gamma_w).$$

If the parameterization "variant=1" is used, similar results can be obtained for fitted values via the transformation

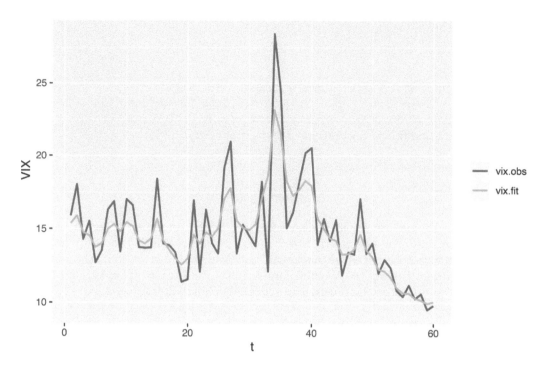

FIGURE 7.6: Observed (red) and fitted (blue) values from the model without regressors (Model W1).

$$E(y_t|\gamma_w, \lambda_t) = \lambda_t^{-1}\Gamma(1 + 1/\gamma_w).$$

Model W2

The Weibull analog of Model G2 (model with VXN as regressor with a dynamic coefficient) is shown below. For illustrative purposes, we ran this model using **variant=1**.

```
id.vxn <- 1:length(vxn)
vix.data <- cbind.data.frame(vix.data, id.vxn, vxn)
formula.vix.W2 = vix ~ 1 + f(id.vxn,vxn, model = "ar1")
model.vix.W2 = inla(
  formula.vix.W2,
  data = vix.data,
  family = "weibull",
  control.family = list(variant = 1),
  control.predictor = list(compute = TRUE, link = 1),
  control.compute = list(
    dic = TRUE,
    waic = TRUE,
    cpo = TRUE,
    config = TRUE
  )
)
# summary(model.vix.W2)
```

```
format.inla.out(model.vix.W2$summary.fixed[, c(1:5)])
##    name       mean   sd    0.025q  0.5q   0.975q
## 1 Intercept -1.833 0.061 -1.956 -1.832 -1.714
format.inla.out(model.vix.W2$summary.hyperpar[, c(1:2)])
##    name                          mean      sd
## 1 alpha parameter for weibull   25.050    3.866
## 2 Precision for id.vxn        1533.827  891.956
## 3 Rho for id.vxn                 0.992    0.006
```

7.3.1   Forecasting from Weibull models

As in the computation of fitted values, evaluation of forecasts from the Weibull state space model requires use of the mean of the Weibull model with scale parameter λ_t and shape parameter γ_w. For both parameterizations, INLA uses the linear predictor $\eta_t = \log(\lambda_t)$.

For example, in Model W1 we have

$$
\begin{aligned}
Y_t|\lambda_t, \gamma_w &\sim \text{Weibull}(\lambda_t, \gamma_w), \\
\eta_t = \log(\lambda_t) &= \alpha + x_t, \\
x_t &= \phi x_{t-1} + w_t.
\end{aligned}
$$ (7.8)

It follows from the above that

$$
x_t|x_{t-1}, \phi, \sigma_w^2 \sim N(\phi x_{t-1}, \sigma_w^2).
$$

After observing data from n periods, i.e., $\boldsymbol{y}^n = (y_1, \ldots, y_n)$, we have posterior distributions of α, σ_w^2, ϕ and x_n. To predict the next observation y_{n+1}, we need the posterior mean of y_{n+1} given $\boldsymbol{y}^n$, i.e.,

$$
E(y_{n+1}|\boldsymbol{y}^n) = E\Big[E(y_{n+1}|\alpha, \gamma_w, \sigma_w^2, \phi, x_{n+1})\Big],
$$

where the outside expectation is taken with respect to the joint posterior distribution of α, γ_w, σ_w^2, ϕ and x_{n+1}, and

$$
E(y_{n+1}|\alpha, \sigma_w^2, \phi, x_{n+1}) = \begin{cases} \lambda_{n+1}^{-1/\gamma_w} \Gamma(1 + 1/\gamma_w), & \text{for variant} = 0 \\ \lambda_{n+1}^{-1} \Gamma(1 + 1/\gamma_w), & \text{for variant} = 1. \end{cases}
$$

We can evaluate $E(y_{n+1}|\boldsymbol{y}^n)$ by a Monte Carlo average based on M samples from the posterior distribution of α, γ_w, σ_w^2, ϕ, and x_{n+1}, for large M. Specifically, for each realization of (x_n, ϕ, σ_w^2), we draw samples of x_{n+1} from $N(\phi x_n, \sigma_w^2)$, and for each realization of (x_{n+1}, α), we obtain samples of $\eta_{n+1} = \alpha + x_{n+1}$ and $\lambda_{n+1} = e^{\eta_{n+1}}$. Finally, for each realization of $(\lambda_{n+1}, \gamma_w)$ we evaluate $E(y_{n+1}|\alpha, \sigma_w^2, \phi, x_{n+1})$, and use the Monte Carlo average as our prediction. Similarly, predictions for $y_{n+2}, \ldots, y_{n+L}$ will require use of the conditional distribution $x_t|x_{t-1}, \phi, \sigma_w^2$ for $t = n+2, \ldots, n+L$ in a sequential manner. We show the code for forecasting under Model W1.

```
mean <-
  post.sample.W1[[1]]$hyperpar['Rho for id.x'] *
  post.sample.W1[[1]]$latent[2 * n]
sigma2.w <- 1 / post.sample.W1[[1]]$hyperpar['Precision for id.x']
```

```
x.np1 <- rnorm(500, mean = mean, sd = sqrt(sigma2.w))
eta.np1 <- c(tail(post.sample.W1[[1]]$latent, 1)) + x.np1
lambda.np1 <- exp(eta.np1)
gamma.w <-
  post.sample.W1[[1]]$hyperpar['alpha parameter for weibull']
y.np1 <- gamma(1 + 1 / gamma.w) /
  (lambda.np1 ^ (1 / gamma.w))
mean(y.np1)
## [1] 12.17
```

7.4   Beta state space model

Let $Y_t \sim \text{Beta}(\tau\gamma_t, \tau(1-\gamma_t))$, where $0 < \gamma_t < 1$. Note that $E(Y_t|\gamma_t) = \gamma_t$, and $Var(Y_t|\gamma_t, \tau) = \gamma_t(1 - \gamma_t)/(\tau + 1)$, implying that τ is a precision parameter. We assume that

$$\eta_t = \log(\gamma_t/(1 - \gamma_t)) = \alpha + \boldsymbol{f}'_t\boldsymbol{x}_t,$$
$$\boldsymbol{x}_t = \boldsymbol{G}\boldsymbol{x}_{t-1} + \boldsymbol{w}_t, \tag{7.9}$$

where α is an intercept term, $\boldsymbol{x}_t$ is an h-dimensional state vector, $\boldsymbol{f}_t$ is an h-dimensional covariate vector, $\boldsymbol{G}$ is an $h \times h$ transition matrix, and $\boldsymbol{w}_t \sim N(\boldsymbol{0}, \boldsymbol{W})$.

7.4.1   Example: Crest market share

We consider weekly market share data for Crest used by Wichern and Jones (1977) for intervention analysis. Historically, Crest toothpaste was introduced nationally in 1956 when Colgate was enjoying a significant portion of the market share. This dominance of Colgate continued until 1960, and Crest's market share during this period was relatively stable, but in August 1, 1960, the American Dental Association (ADA) endorsed Crest as an "important aid in any program of dental hygiene." In what follows, we first show a time series plot of Crest's market share in Figure 7.7, where the ADA endorsement date of August 1, 1960, is presented by a vertical dashed line at $t = 135$. As shown in this plot, there is a clear jump in Crest's market following the endorsement, suggesting nonstationarity in the data.

```
mshare <- read.csv("TPASTE.csv", header = TRUE)
crestms <- mshare[, 1]
tsline(crestms,
       ylab = "market share",
       xlab = "week",
       line.color = "red",
       line.size = 0.6) +
  geom_vline(xintercept = 135, linetype="dashed", color = "black")

# abline(v = 135, col = "blue", lty = 2)
```

As in the previous section, if we do not have any regressors, then $\eta_t = \alpha + x_t$ will be a scalar, and the state equation will reduce to

$$x_t = Gx_{t-1} + w_t, \tag{7.10}$$

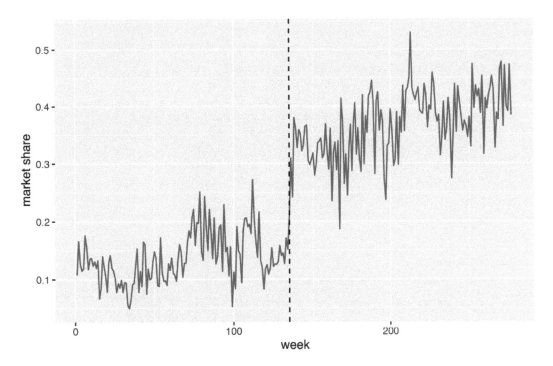

FIGURE 7.7: Weekly market share of Crest toothpaste. The vertical line (black) denotes Aug 1, 1960.

where G is a scalar and and w_t is a univariate white noise process.

Model B1

We first consider the model without any regressors, i.e.,

$$\eta_t = \alpha + x_t,$$
$$x_t = \phi x_{t-1} + w_t. \tag{7.11}$$

We estimate the model using the code below and show the plot of fitted versus the observed values in Figure 7.8.

```
id.x <- 1:length(crestms)
crest.data <- cbind.data.frame(mshare, id.x)
formula.crest.B1 = crestms ~ 1 + f(id.x, model = "ar1")
model.crest.B1 = inla(
  formula.crest.B1,
  data = crest.data,
  family = "beta",
  scale = 1,
  control.predictor = list(compute = TRUE),
  control.compute = list(
    dic = TRUE,
    waic = TRUE,
    cpo = TRUE,
```

```
    config = TRUE
  )
)
# summary(model.crest.B1)
format.inla.out(model.crest.B1$summary.fixed[, c(1:5)])
##    name        mean    sd    0.025q  0.5q    0.975q
## 1 Intercept  -1.159  0.46  -2.106  -1.162  -0.191
format.inla.out(model.crest.B1$summary.hyperpar[, c(1:2)])
##    name                                mean    sd
## 1 precision parameter for beta obs    114.87  11.897
## 2 Precision for id.x                   2.56   1.236
## 3 Rho for id.x                         0.99   0.005
```

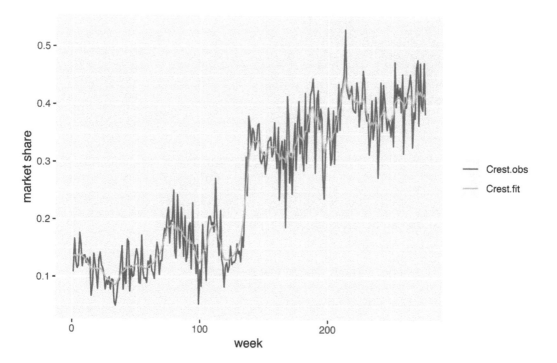

FIGURE 7.8: Observed (red) and fitted (blue) values from Model B1 for the Crest data.

Model B2

Next, we bring in three regressors, the market share for Colgate (*colgms*), unit price for Crest (crestpr), and unit price for Colgate (*colgpr*). We treat them all as fixed, but include a dynamic intercept term, i.e.,

$$
\begin{aligned}
\eta_t &= \alpha + \beta_1 Z_{1t} + \beta_2 Z_{2t} + \beta_3 Z_{3t} + x_t, \\
x_t &= \phi x_{t-1} + w_t.
\end{aligned} \tag{7.12}
$$

The plot of the actual versus fitted series is shown in Figure 7.9.

```
colgms <- mshare[, 2]
crestpr <- mshare[, 3]
colgpr <- mshare[, 4]
formula.crest.B2 <-
  crestms ~ 1 + colgms + crestpr + colgpr + f(id.x, model = "ar1")
model.crest.B2 <- inla(
  formula.crest.B2,
  data = crest.data,
  family = "beta",
  scale = 1,
  control.predictor = list(compute = TRUE),
  control.compute = list(
    dic = TRUE,
    waic = TRUE,
    cpo = TRUE,
    config = TRUE
  )
)
# summary(model.crest.B2)
format.inla.out(model.crest.B2$summary.fixed[, c(1:5)])
##    name        mean    sd    0.025q  0.5q    0.975q
## 1 Intercept    2.371  0.779  0.818   2.378   3.881
## 2 colgms      -2.593  0.315 -3.212  -2.593  -1.976
## 3 crestpr     -0.887  0.241 -1.360  -0.887  -0.413
## 4 colgpr      -0.515  0.286 -1.077  -0.516   0.047
format.inla.out(model.crest.B2$summary.hyperpar[, c(1:2)])
##    name                                   mean     sd
## 1 precision parameter for beta obs       135.576  13.737
## 2 Precision for id.x                       6.954   3.184
## 3 Rho for id.x                             0.985   0.007
```

The signs for the coefficients of *colgms* and *crestpr* are negative, as expected. While the sign of *colgpr* being negative is not as easy to explain, we note that this coefficient is not significantly different than zero.

Model B3

Our third model will have dynamic coefficients for all three covariates. Specifically, we have

$$
\begin{aligned}
\eta_t &= \alpha + \beta_{1t} Z_{1t} + \beta_{2t} Z_{2t} + \beta_{3t} Z_{3t}, \\
\beta_{1t} &= \phi_1 \beta_{1,t-1} + w_{1t}, \\
\beta_{2t} &= \phi_2 \beta_{2,t-1} + w_{2t}, \\
\beta_{3t} &= \phi_3 \beta_{3,t-1} + w_{3t}.
\end{aligned} \tag{7.13}
$$

We show a plot of actual versus fitted observations in Figure 7.10.

```
id.colgms <- id.crestpr <- id.colgpr <- 1:length(crestms)
formula.crest.B3 <-
  crestms ~ 1 + f(id.colgms, colgms, model = "ar1") +
  f(id.crestpr, crestpr, model = "ar1") +
  f(id.colgpr, colgpr, model = "ar1")
model.crest.B3 <- inla(
```

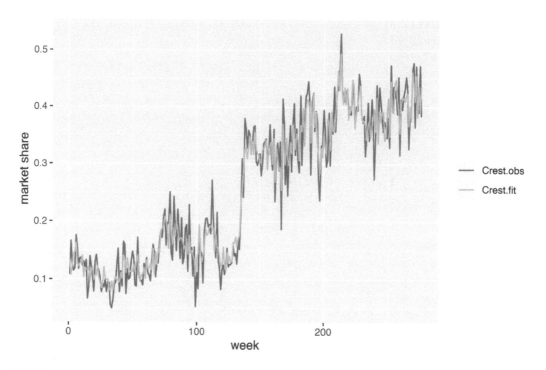

FIGURE 7.9: Observed (red) and fitted (blue) values from Model B2 for the Crest data.

```
    formula.crest.B3,
    data = crest.data,
    family = "beta",
    scale = 1,
    control.predictor = list(compute = TRUE),
    control.compute = list(
      dic = TRUE,
      waic = TRUE,
      cpo = TRUE,
      config = TRUE
    )
  )
)
# summary(model.crest.B3)
format.inla.out(model.crest.B3$summary.fixed[, c(1:5)])
##    name       mean  sd    0.025q 0.5q  0.975q
## 1 Intercept 0.138 0.527 -0.837 0.129 1.191
format.inla.out(model.crest.B3$summary.hyperpar[, c(1:2)])
##    name                             mean       sd
## 1 precision parameter for beta obs  116.106    12.072
## 2 Precision for id.colgms           22971.649 23826.284
## 3 Rho for id.colgms                 0.031      0.674
## 4 Precision for id.crestpr          5.214      3.632
## 5 Rho for id.crestpr                0.995      0.003
```

```
## 6 Precision for id.colgpr        22648.905 21768.024
## 7 Rho for id.colgpr                  0.047      0.674
```

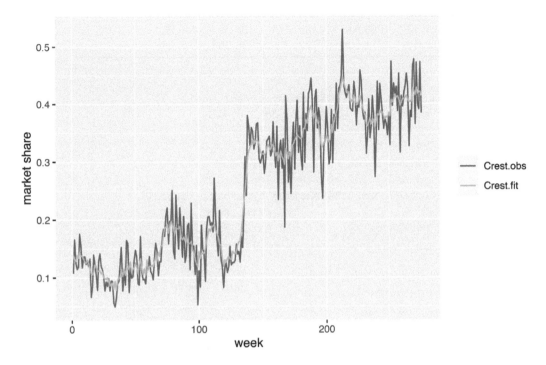

FIGURE 7.10: Observed (red) and fitted (blue) values from Model B3 for the Crest data.

Model B4

We can also try fitting a standard static beta regression model as a benchmark, i.e.,

$$\eta_t = \alpha + \beta_1 Z_{1t} + \beta_2 Z_{2t} + \beta_3 Z_{3t}. \qquad (7.14)$$

The fit of the model is shown in Figure 7.11.

```
formula.crest.B4 <- crestms ~ 1 + colgms + crestpr + colgpr
model.crest.B4 <- inla(
  formula.crest.B4,
  data = crest.data,
  family = "beta",
  scale = 1,
  control.predictor = list(compute = TRUE),
  control.compute = list(
    dic = TRUE,
    waic = TRUE,
    cpo = TRUE,
    config = TRUE
  )
)
# summary(model.crest.B4)
```

TABLE 7.2: Comparison of Models B1, B2, B3, and B4.

Model	DIC	WAIC	PsBF	Marginal Likelihood
Model B1	−994.35	−996.223	495.744	446.508
Model B2	−1052.07	−1054.711	525.875	469.797
Model B3	−998.985	−1001.029	498.292	449.91
Model B4	−771.608	−772.073	386.034	368.948

```
format.inla.out(model.crest.B4$summary.fixed[, c(1:5)])
##   name       mean    sd     0.025q  0.5q   0.975q
## 1 Intercept  6.285  0.395   5.509   6.286  7.059
## 2 colgms     -4.196 0.411  -5.004  -4.196 -3.390
## 3 crestpr    -2.649 0.245  -3.130  -2.649 -2.168
## 4 colgpr     -0.485 0.252  -0.979  -0.485  0.009
format.inla.out(model.crest.B4$summary.hyperpar[, c(1:2)])
##   name                              mean   sd
## 1 precision parameter for beta obs 44.45  3.763
```

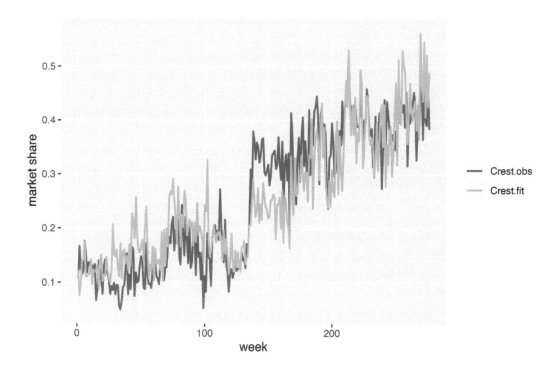

FIGURE 7.11: Observed and fitted values from Model B4 for the Crest data.

We present a comparison of all the models in Table 7.2. Note that among the four models, Model B2, where we have static regression coefficients and a dynamic intercept, gives the best fit.

8

Modeling Categorical Time Series

8.1 Introduction

This chapter describes Bayesian dynamic generalized linear models (DGLMs) for categorical time series. Section 8.2 discusses model fitting for a single binomial response time series, while Section 8.3 describes a hierarchical framework for modeling multiple binomial response time series (see Chapter 6 for a discussion on *Gaussian* hierarchical dynamic models). Modeling time series with more than two categories is discussed in Section 8.4.

The custom functions must first be sourced.

```
source("functions_custom.R")
```

We use the following packages in this chapter.

```
library(INLA)
library(tidyverse)
```

We analyze the following data set.

1. Weekly shopping trips

8.2 Binomial response time series

Consider a discrete-valued time series Y_t whose conditional distribution at time t is $\text{Bin}(m, \pi_t)$, where m denotes the size of the distribution. Then, Y_t denotes the number of successes in m independent Bernoulli trials, and has mean and variance given by $E(Y_t) = m\pi_t$ and $\text{Var}(Y_t) = m\pi_t(1 - \pi_t)$. Following the well known generalized linear modeling approach (McCullagh and Nelder, 1989), we can link the mean response π_t to a structured linear Gaussian predictor using a suitable link function. Well known link functions for binomial response modeling include the logit and probit links for π_t, given by

$$\text{logit}(\pi_t) = \log \frac{\pi_t}{1 - \pi_t},$$

and

$$\text{probit}(\pi_t) = \mathcal{N}^{-1}(\pi_t),$$

where $\mathcal{N}^{-1}(.)$ denotes the inverse c.d.f. of the $N(0, 1)$ distribution. Note that we have used $\mathcal{N}^{-1}$ instead of the usual notation in the literature of Φ^{-1} to avoid confusion with the matrix of AR coefficients (see Chapter 12). Excellent references on sampling based Bayesian approaches for fitting DGLMs include Gamerman (1998) and West and Harrison (1997). Also, see Soyer and Sung (2013) for a discussion of binary longitudinal response modeling under a probit link.

8.2.1   Example: Simulated single binomial response series

The observation and state equations of a DGLM for a binomial response time series Y_t with a logit link are shown below:

$$Y_t|\pi_t \sim \text{Binomial}(m, \pi_t),$$
$$\text{logit}(\pi_t) = \alpha + \beta_{t,0},$$
$$\beta_{t,0} = \phi\beta_{t-1,0} + w_{t,0}. \tag{8.1}$$

The model assumes a time-varying intercept $\beta_{t,0}$ following a latent Gaussian AR(1) model to explain $\text{logit}(\pi_t)$, where $|\phi| < 1$ and the state error $w_{t,0}$ is $N(0, \sigma_w^2)$. We have also included a fixed level α.

We show the model setup on a simulated time series of length $n = 1000$, with a burn-in of length 500. The size m of the binomial distribution is denoted by *bin.size*, and will be passed to **inla()** through the argument *Ntrials*.

```
set.seed(123457)
n <- 1000
burn.in <- 500
n.tot <- n + burn.in
bin.size <- 10
sigma2.w <- 0.05
phi <- 0.8
alpha <- 1.3
yit <- beta.t0 <- pi.t <- eta.t <- vector("double", n.tot)
beta.t0 <-
  arima.sim(list(ar = phi, ma = 0), n = n.tot, sd = sqrt(sigma2.w))
for (i in 1:n.tot) {
  eta.t[i] <- alpha + beta.t0[i]
  pi.t[i] <- exp(eta.t[i]) / (1 + exp(eta.t[i]))
  yit[i] <- rbinom(n = 1, size = bin.size, prob = pi.t[i])
}
yit <- tail(yit, n)
```

The true precision of the state error $w_{t,0}$ is 20, while the true precision of the state variable $\beta_{t,0}$ is 7.2, which can be computed using the book's custom function **prec.state.ar1()**:

```
prec.state.ar1(state.sigma2 = sigma2.w, phi = phi)
```

We fit the model in (8.1) to the simulated data.

```
id.beta0 <- 1:n
data.bin <- cbind.data.frame(id.beta0, yit)
prior.bin <- list(
  prec = list(prior = "loggamma", param = c(0.1, 0.1)),
  rho = list(prior = "normal", param = c(0, 0.15))
)
```

```
formula.bin <-
  yit ~ 1 + f(id.beta0,
              model = "ar1",
              constr = FALSE,
              hyper = prior.bin)
model.bin <- inla(
  formula.bin,
  family = "binomial",
  Ntrials = bin.size,
  control.family = list(link = "logit"),
  data = data.bin,
  control.predictor = list(compute = TRUE, link = 1),
  control.compute = list(
    dic = TRUE,
    cpo = TRUE,
    waic = TRUE,
    config = TRUE,
    cpo = TRUE
  )
)
# summary(model.bin)
format.inla.out(model.bin$summary.fixed[,c(1,2,4)])
## name       mean  sd    0.5q
## 1 Intercept 1.228 0.044 1.227
format.inla.out(model.bin$summary.hyperpar[, c(1,2,4)])
## name                       mean  sd    0.5q
## 1 Precision for id.beta0 8.773 2.019 8.533
## 2 Rho for id.beta0       0.818 0.056 0.825
```

The posterior means for α, ϕ and σ_w^2 are 1.228, 0.818 and 8.773 respectively, suggesting that the true parameters in the simulation are recovered well.

Note that if we *incorrectly* fit a model with no level α to this data, then neither the posterior mean of ϕ nor the posterior mean of the precision of the state variable are recovered well. In particular, the posterior mean of ϕ increases to 0.996 (i.e., it tends to a limit of 1), possibly in order to accommodate the omitted non-zero level. We illustrate this below. We reiterate what we mentioned in Section 3.3, that it is advisable to include a level α in the model, expecting it to be estimated close to zero in cases when the true level is zero.

```
formula.bin.noalpha <-
  yit ~ -1 + f(id.beta0,
               model = "ar1",
               constr = FALSE)
model.bin.noalpha <- inla(
  formula.bin.noalpha,
  family = "binomial",
  Ntrials = bin.size,
  control.family = list(link = "logit"),
  data = data.bin,
  control.predictor = list(compute = TRUE, link = 1),
  control.compute = list(
    dic = TRUE,
```

```
    cpo = TRUE,
    waic = TRUE,
    config = TRUE,
    cpo = TRUE
  )
)
format.inla.out(model.bin.noalpha$summary.hyperpar[, c(1,2,4)])
##    name                    mean  sd    0.5q
## 1 Precision for id.beta0 1.144 0.451 1.075
## 2 Rho for id.beta0        0.996 0.002 0.996
```

8.2.2   Example: Weekly shopping trips for a single household

The dataset *trips* consists of the number of weekly shopping trips for $N = 548$ households across $n = 104$ weeks (Soyer et al., 2016). Here, we present our analysis for a *single household*, while we discuss a model for all households using a hierarchical model in Section 8.3.2. In this example, we set m or *bin.size* to be the maximum number of shopping trips taken by this household over the 104 weeks, and assume that the observed weekly shopping trips Y_t has a $\text{Bin}(m, \pi_t)$ distribution.

```
trips.single.household <-
  read.csv("weekly_shopping_single_household.csv",
           header = TRUE,
           stringsAsFactors = FALSE)
tsline(trips.single.household$n.trips,
       xlab = "week",
       ylab = "shopping trips",
       line.color = "red",
       line.size = 0.6)
```

Figure 8.1 shows a plot of Y_t for a household with ID *14100859*. We divide the data into training and test portions, and create a data frame which includes an index id.beta0.

```
n <- nrow(trips.single.household)
n.hold <- 6
n.train <- n - n.hold
train.trips.single.household <-
  trips.single.household$n.trips[1:n.train]
test.trips.single.household <-
  tail(trips.single.household$n.trips, n.hold)
bin.size <- max(train.trips.single.household)
id.beta0 <- 1:n
data.single.household <-
  cbind.data.frame(n.trips = c(train.trips.single.household,
                               rep(NA, n.hold)), id.beta0)
```

To model the time series of trips using (8.1), we use a non-default prior specification for $1/\sigma_w^2$; specifically, we assume small values of 0.1 and 0.1 for the shape and scale parameters of the log gamma distribution, leading to a diffuse prior. The AR(1) parameter has the default prior for its internal parametrization, as discussed in Chapter 3.

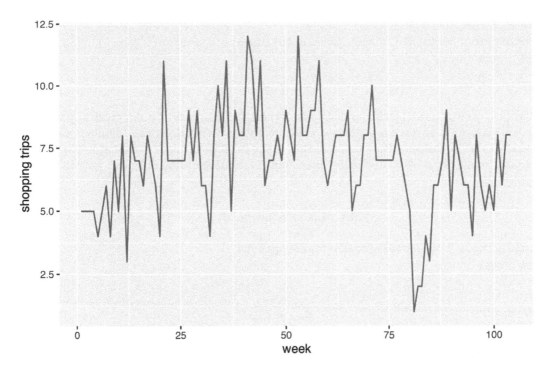

FIGURE 8.1: Number of weekly shopping trips for a single household over n = 104 weeks.

```
prior.single.household <- list(
  prec = list(prior = "loggamma", param = c(0.1, 0.1)),
  rho = list(prior = "normal", param = c(0, 1))
)
```

The formula and results are shown below.

```
formula.single.household <-
  n.trips ~ 1 + f(id.beta0, model = "ar1", hyper = prior.single.household)
inla.single.household <- inla(
  formula.single.household,
  family = 'binomial',
  Ntrials = bin.size,
  data = data.single.household,
  control.family = list(link = 'logit'),
  control.predictor = list(link = 1, compute = TRUE),
  control.compute = list(
    dic = TRUE,
    cpo = TRUE,
    waic = TRUE,
    config = TRUE
  )
)
# summary(inla.single.household)
format.inla.out(inla.single.household$summary.fixed[, c(1, 3, 5)])
```

```
##    name        mean   0.025q  0.975q
## 1 Intercept  0.288  -0.11   0.652
format.inla.out(inla.single.household$summary.hyperpar[, c(1:2)])
##    name                     mean   sd
## 1 Precision for id.beta0  4.625  1.947
## 2 Rho for id.beta0         0.802  0.089
```

The posterior mean and standard deviation of the hyperparameter ϕ are respectively 0.802 and 0.089, showing that there is a strong positive temporal correlation in the number of shopping trips for this household. We obtain forecasts using `inla.posterior.sample()`. The contents of each sample can be viewed using `inla.single.household$misc$configs$contents`.

```
post.samp <-
  inla.posterior.sample(n = 1000,
                        inla.single.household,
                        seed = 1234)
inla.single.household$misc$configs$contents
## $tag
## [1] "Predictor"   "id.beta0"    "(Intercept)"
##
## $start
## [1]    1 105 209
##
## $length
## [1] 104 104    1
```

Here "Predictor" returns the predicted η_t for each t. All the elements in `tag` are stored under *latent* for each posterior sample. The first n rows are the "Predictors", the next n rows are the posterior estimates of "id.beta0" and the last row contains the posterior estimate of "(Intercept)". Here, we only show the first and last 6 rows of *latent* for the first sample.

```
head(post.samp[[1]]$latent)
##                  sample:1
## Predictor:1  -0.18534
## Predictor:2  -0.15249
## Predictor:3  -0.20560
## Predictor:4   0.28900
## Predictor:5   0.02523
## Predictor:6  -0.39242
tail(post.samp[[1]]$latent)
##                  sample:1
## id.beta0:100    0.1207
## id.beta0:101    0.7823
## id.beta0:102    0.8362
## id.beta0:103    0.6775
## id.beta0:104    0.1225
## (Intercept):1   0.2945
```

We transform the predicted $\eta_{j,t}$ into the predicted $\tilde{\pi}_{j,t}$ via

$$\tilde{\pi}_{j,t} = \frac{\exp(\eta_{j,t})}{1 + \exp(\eta_{j,t})},$$

for each Monte Carlo realization $j = 1, \ldots, 1000$. For each j, we use `rbinom()` to simulate a fitted response from $\text{Bin}(m, \tilde{\pi}_{j,t})$ for $t = 1, \ldots, n$, yielding 1000 binomial time series. We can then obtain summaries such as the mean, standard deviation, and percentiles from these posterior predictive distributions, as shown below.

```r
fit.post.samp <- vector("list", 1000)
for (j in 1:1000) {
  eta <- post.samp[[j]]$latent[1:n]
  p <- exp(eta) / (1 + exp(eta))
  fit.post.samp[[j]] <- rbinom(n, bin.size, p)
}
fore.mean <- bind_cols(fit.post.samp) %>%
  apply(1, mean)
fore.sd <- bind_cols(fit.post.samp) %>%
  apply(1, sd) %>%
  round(3)
fore.quant <- bind_cols(fit.post.samp) %>%
  apply(1, function(x)
    quantile(x, probs = c(0.025, 0.975)))
obs.fit.single.household <-
  cbind.data.frame(
    observed = trips.single.household$n.trips,
    fore.mean,
    fore.sd,
    quant0.025 = fore.quant[1,],
    quant0.975 = fore.quant[2,]
  )
```

Figure 8.2 shows the observed weekly shopping trips along with the mean of the forecast values generated from `inla.posterior.sample()`.

For this household, the fixed level α is not significant, and fitting the model without a level α yields results that are quite similar to what we have shown, as we see below.

```r
formula.single.household.noalpha <-
  n.trips ~ -1 + f(id.beta0,  model = "ar1", hyper = prior.single.household)
inla.single.household.noalpha <- inla(
  formula.single.household.noalpha,
  family = 'binomial',
  Ntrials = bin.size,
  data = data.single.household,
  control.family = list(link = 'logit'),
  control.predictor = list(link = 1, compute = TRUE),
  control.compute = list(dic = TRUE, cpo = TRUE, waic =
                         TRUE)
)
format.inla.out(inla.single.household.noalpha$summary.hyperpar[, c(1:2)])
##    name                  mean   sd
## 1 Precision for id.beta0 3.947 1.571
## 2 Rho for id.beta0       0.844 0.066
```

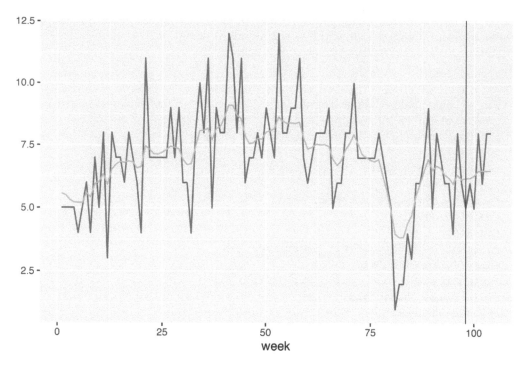

FIGURE 8.2: Observed (red) versus fitted values (blue) for weekly shopping trips of a single household. The black vertical line splits the data into training and test portions.

8.3 Modeling multiple binomial response time series

In many situations, we may wish to model several binomial response time series Y_{it} for $t = 1, \ldots, n$ weeks and $i = 1, \ldots, g$ households by aggregating information from all the series. Assume that Y_{it} follows a Binomial(m_i, π_{it}) distribution with mean $E(Y_{it}) = m_i \pi_{it}$ and variance Var($Y_{it}) = m_i \pi_{it}(1 - \pi_{it})$. Again, each π_{it} is linked to a structured predictor which may be a function of exogenous predictors, or suitable functions of time, or both. In a dynamic model, the state variables may be indexed by i, or t, or both. The coefficients corresponding to the exogenous predictors may be static or dynamic. They can also be constant across i or vary by i. A similar approach was described in Chapter 6, where we discussed hierarchical DLMs for g continuous-valued response time series.

8.3.1 Example: Dynamic aggregated model for multiple binomial response time series

Consider the observation and state equations for a hierarchical DGLM (HDGLM) for g binomial response series under a logit link:

$$Y_{it}|\pi_{it} \sim \text{Bin}(m_i, \pi_{it}),$$
$$\text{logit}(\pi_{it}) = \alpha + \beta_{it,0} + \beta_{it,1}Z_{it,1} + \beta_2 Z_{it,2},$$
$$\beta_{it,0} = \phi_0\beta_{i,t-1,0} + w_{it,0},$$
$$\beta_{it,1} = \phi_1\beta_{i,t-1,1} + w_{it,1}. \tag{8.2}$$

Here, α is a fixed level, while the intercept $\beta_{it,0}$ varies over both t and i; $\beta_{it,1}$ is a random effect corresponding to an exogenous predictor $Z_{it,1}$, and we assume that it varies over both t and i; and β_2 is a static (fixed) effect corresponding to another exogenous predictor $Z_{it,2}$. Specifically, $\beta_{it,0}$ follows a latent Gaussian AR(1) process with the same coefficient ϕ_0 across all i, where $|\phi_0| < 1$, and state error $w_{it,0}$ which is $N(0, \sigma_{w0}^2)$. Likewise, $\beta_{it,1}$ is a latent Gaussian AR(1) process with coefficient ϕ_1 (with $|\phi_1| < 1$), and state error $w_{it,1} \sim N(0, \sigma_{w1}^2)$.

Values of the sizes m_i, $i = 1, \ldots, g$, are randomly selected from the set of integers $\{1, 2, \ldots, 12\}$, and the true parameter values are shown below.

```
set.seed(123457)
g <- 300
n <- 100
burn.in <- 100
n.tot <- n + burn.in
bin.size <- sample(1:12, size = g, replace = TRUE)
alpha <- 1.3
beta.2 <- 2.0
sigma2.w0 <- 0.15
sigma2.w1 <- 0.05
phi.0 <- 0.8
phi.1 <- 0.4
```

We simulate $g = 300$ binomial response time series from (8.2), each of length $n = 100$ as follows.

```
yit.all <- z1.all <- z2.all <- vector("list", g)
for (j in 1:g) {
  z1 <- rnorm(n.tot, 0, 1)
  z2 <- rnorm(n.tot, 0, 1)
  yit <-
    beta.0t <- beta.1t <- pi.t <- eta.t <- vector("double", n.tot)
  beta.0t <-
    arima.sim(list(ar = phi.0, ma = 0),
              n = n.tot,
              sd = sqrt(sigma2.w0))
  beta.1t <-
    arima.sim(list(ar = phi.1, ma = 0),
              n = n.tot,
              sd = sqrt(sigma2.w1))
  for (i in 1:n.tot) {
    eta.t[i] <- alpha + beta.0t[i] + beta.1t[i] * z1[i] + beta.2 * z2[i]
    pi.t[i] <- exp(eta.t[i]) / (1 + exp(eta.t[i]))
    yit[i] <- rbinom(n = 1,
                     size = bin.size[j],
```

```
                      prob = pi.t[i])
  }
  yit.all[[j]] <- tail(yit, n)
  z1.all[[j]] <- tail(z1, n)
  z2.all[[j]] <- tail(z2, n)
}
yit.all <- unlist(yit.all)
z1.all <- unlist(z1.all)
z2.all <- unlist(z2.all)
```

Recall from Chapter 6, that to set up a hierarchical model, it is necessary to define appropriate indexes like *id.beta0* and *id.beta1* to capture the time varying effect and *re.beta0* and *re.beta1* to allow the intercept and covariate $Z_{it,1}$ to vary over groups $i = 1, \ldots, g$. In the binomial response models, note that the size *Ntrials* must be provided for each i via `bin.size.all <- rep(bin.size, each = n)`.

```
id.beta0 <- id.beta1 <- rep(1:n, g)
re.beta0 <- re.beta1 <- rep(1:g, each = n)
bin.size.all <- rep(bin.size, each = n)
data.simul.hbin <-
  cbind.data.frame(id.beta0,
                   id.beta1,
                   yit.all,
                   z1.all,
                   re.beta0,
                   re.beta1,
                   bin.size.all)
```

We use non-default priors for ϕ and σ_w^2 (in their internal parametrizations).

```
prior.hbin <- list(
  prec = list(prior = "loggamma", param = c(0.1, 0.1)),
  rho = list(prior = "normal", param = c(0, 0.15))
)
```

The formula and `inla()` call are shown below, followed by results from fitting the model in (8.2) to this data. The fitted model for the aggregated binomial responses recovers the true parameters well, as seen from the posterior summaries of the fixed effects and the hyperparameters.

```
formula.simul.hbin <-
  yit.all ~ 1 + z2.all +
  f(id.beta0,
    model = "ar1",
    replicate = re.beta0,
    hyper = prior.hbin) +
  f(
    id.beta1,
    z1.all,
    model = "ar1",
    replicate = re.beta1,
    hyper = prior.hbin
  )
```

```
model.simul.hbin <- inla(
  formula.simul.hbin,
  family = "binomial",
  Ntrials = bin.size.all,
  control.family = list(link = "logit"),
  data = data.simul.hbin,
  control.predictor = list(compute = TRUE, link = 1),
  control.compute = list(
    dic = TRUE,
    cpo = TRUE,
    waic = TRUE,
    config = TRUE,
    cpo = TRUE
  )
)
```

```
format.inla.out(summary.simul.hbin$fixed[, c(1:5)])
##    name         mean   sd     0.025q  0.5q   0.975q
## 1 Intercept    1.279  0.014  1.252   1.279  1.306
## 2 z2.all       1.978  0.011  1.956   1.978  2.000
format.inla.out(summary.simul.hbin$hyperpar[, c(1:2)])
##    name                          mean    sd
## 1 Precision for id.beta0         2.600   0.098
## 2 Rho for id.beta0               0.811   0.009
## 3 Precision for id.beta1        20.046   3.357
## 4 Rho for id.beta1               0.263   0.324
```

8.3.2 Example: Weekly shopping trips for multiple households

We model the number of weekly shopping trips, i.e., *n.trips*, of $g = 548$ households over $n = 104$ weeks (Soyer et al., 2016). We fit a hierarchical model to the binomial responses Y_{it}. The size m_i of the binomial distribution for each household is taken to be the maximum of the shopping trips across all weeks for that household. The model has a random household-specific and time-varying intercept, but no exogenous predictors.

$$Y_{it}|\pi_{it} \sim \text{Bin}(m_i, \pi_{it}),$$
$$\text{logit}(\pi_{it}) = \alpha + \beta_{it,0},$$
$$\beta_{it,0} = \phi_0 \beta_{i,t-1,0} + w_{it,0}, \tag{8.3}$$

```
trips <-
  read.csv("Weekly_Shopping_Trips.csv",
           header = TRUE,
           stringsAsFactors = FALSE)
```

Each row of the data set *trips* contains the weekly number of trips for each of $g = 548$ households over $n = 104$ weeks. We first convert data from *wide data* format to *long data* format, which facilitates our model setup.

```
g <- nrow(trips)
trips.long <- trips %>%
  pivot_longer(cols = W614:W717, names_to = "week") %>%
```

```
    rename(n.trips = value)
n <- n_distinct(trips.long$week)
n.hold <- 6
n.train <- n - n.hold
trips.house <- trips.long %>%
  group_split(panelist)
train.df <- trips.house %>%
  lapply(function(x)
    x[1:n.train, ]) %>%
  bind_rows()
test.df <- trips.house %>%
  lapply(function(x)
    tail(x, n.hold)) %>%
  bind_rows()
```

We set up indexes and replicates for fitting (8.3).

```
bin.size <- train.df %>%
  group_split(panelist) %>%
  sapply(function(x)
    max(x$n.trips))
trips.allh <- trips.house %>%
  lapply(function(x)
    mutate(x, n.trips = replace(
      n.trips, list = (n.train + 1):n, values = NA
    )))
id.b0 <- id.house <- rep(1:n, g)
re <- re.house <- rep(1:g, each = n)
trips.allh <-
  cbind.data.frame(bind_rows(trips.allh), id.b0, re, id.house, re.house)
prior.hbin = list(
  prec = list(prior = "loggamma", param = c(0.1, 0.1)),
  rho = list(prior = "normal", param = c(0, 0.15))
)
bin.size.all <- rep(bin.size, each = n)
```

The formula and results are shown below.

```
formula.allh <- n.trips ~ 1 +
  f(id.b0,
    model = "ar1",
    replicate = re,hyper = prior.hbin) +
  f(id.house, model = "iid", replicate = re.house)
model.allh <- inla(
  formula.allh,
  family = "binomial",
  Ntrials = bin.size.all,
  control.family = list(link="logit"),
  data = trips.allh,
  control.predictor = list(compute = TRUE, link = 1),
  control.compute = list(
    dic = TRUE,
```

```
    cpo = TRUE,
    waic = TRUE,
    config = TRUE,
    cpo = TRUE)
)
# summary(model.allh)
```

```
format.inla.out(model.allh$summary.fixed[,c(1:2)])
##    name       mean    sd
## 1 Intercept -0.737 0.019
format.inla.out(model.allh$summary.hyperpar[,c(1:2)])
##    name                   mean    sd
## 1 Precision for id.b0      2.575 0.104
## 2 Rho for id.b0            0.975 0.002
## 3 Precision for id.house 11.480 1.095
```

Similar to Section 8.2.2, we obtain the posterior predictions and their summaries for all g households using `inla.posterior.sample()`.

```
set.seed(1234579)
post.samp <- inla.posterior.sample(n = 1000, model.allh)
model.allh$misc$configs$contents
fit.post.samp <- vector("list", 1000)
fit.house <-
  matrix(0,
         nrow = n,
         ncol = g,
         dimnames = list(NULL, paste("H", 1:g, sep = ".")))
for (j in 1:1000) {
  p <- post.samp[[j]]$latent[1:(n * g)] %>%
    matrix(nrow = n,
           ncol = g,
           byrow = F)
  for (i in 1:g) {
    fit.house[, i] <-
      sapply(p[, i], function(x)
        rbinom(1, bin.size[g], exp(x) / (1 + exp(x))))
  }
  fit.post.samp[[j]] <- fit.house
}
fit.post.samp.all <- fit.post.samp %>%
  lapply(as_tibble) %>%
  bind_rows()
fit.per.house <- vector("list", g)
length(fit.per.house)
str(fit.post.samp.all[, 1])
for (k in 1:g) {
  fit.per.house[[k]] <-
    matrix(c(fit.post.samp.all[[k]]), nrow = n, ncol = 1000)
}
```

In Figure 8.3, we show plots of observed and fitted values from the hierarchical model (8.3) for a random sample of four households.

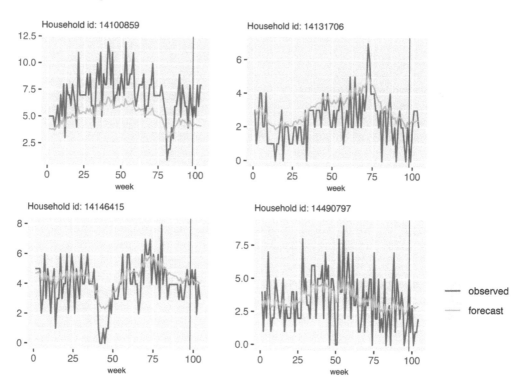

FIGURE 8.3: Plots of observed (red) and fitted (blue) series for four different households.

8.4   Multinomial time series

We end this chapter with a discussion on using R-INLA for modeling categorical time series that follow a multinomial distribution; see Cargnoni et al. (1997). Consider a categorical variable with J mutually exclusive levels. For example, the variable could be the channel for selling a product, with $J = 3$ levels corresponding to brick-and-mortar stores, discount stores, and online sites. Suppose that out of a total of Y_{t+} units at time t, $Y_{j,t}$ units correspond to level j, $j = 1, \ldots, J$, such that $Y_{j,t}$ are non-negative integers, and $\sum_{j=1}^{J} Y_{j,t} = Y_{t+}$. At each time t, given Y_{t+} (i.e., assuming that Y_{t+} is a fixed constant), suppose that the random vector $\boldsymbol{y}_t = (Y_{1,t}, \ldots, Y_{J,t})'$ has a multinomial distribution, $MN(Y_{t+}, \pi_{1,t}, \ldots, \pi_{J,t})$, with p.m.f.

$$
p(\boldsymbol{y}_t; \boldsymbol{\pi}_t, Y_{t+}) = \frac{Y_{t+}!}{\prod_{j=1}^{J} y_{j,t}!} \prod_{j=1}^{J} \pi_{j,t}^{Y_{j,t}}
$$

$$
= \frac{\Gamma(Y_{t+} + 1)}{\prod_{j=1}^{J} \Gamma(Y_{j,t} + 1)} \prod_{j=1}^{J} \pi_{j,t}^{Y_{j,t}}, \tag{8.4}
$$

where $\boldsymbol{\pi}_t = (\pi_{1,t}, \ldots, \pi_{J,t})'$, $0 \leq \pi_{j,t} \leq 1$ and $\sum_{j=1}^{J} \pi_{j,t} = 1$.

At each time t, assume that we have a k-dimensional vector of predictors $\boldsymbol{Z}_{j,t}$, which may also vary by level j. Let $\boldsymbol{Y}^n = (\boldsymbol{y}'_1, \ldots, \boldsymbol{y}'_n)'$ denote the observed multinomial data over n time periods (i.e., $t = 1, \ldots, n$). Assume that the proportions $\pi_{j,t}$ are proportional to the regression function $g_{j,t}(\boldsymbol{\beta}, \boldsymbol{Z}_{j,t})$, $j = 1, \ldots, J$, where $\boldsymbol{\beta}$ represents the k-dimensional vector of regression coefficients. Let $\eta_{j,t} = \boldsymbol{Z}'_{j,t}\boldsymbol{\beta}$. For simplicity of notation, denote $g_{j,t}(\boldsymbol{\beta}, \boldsymbol{Z}_{j,t})$ by $g_{j,t}(\boldsymbol{\beta})$, and let $G_t(\boldsymbol{\beta}) = \sum_{j=1}^{J} g_{j,t}(\boldsymbol{\beta})$. The most commonly used link function specifies

$$g_{j,t}(\boldsymbol{\beta}) = \exp(\eta_{j,t}) = \exp(\boldsymbol{z}'_{j,t}\boldsymbol{\beta}). \qquad (8.5)$$

The multinomial proportions are expressed as functions of $\boldsymbol{\beta}$, i.e.,

$$\pi_{j,t}(\boldsymbol{\beta}) = \frac{g_{j,t}(\boldsymbol{\beta})}{G_t(\boldsymbol{\beta})}, \qquad (8.6)$$

and the multinomial likelihood can be written as

$$L_{\mathcal{M}}(\boldsymbol{\beta}; \boldsymbol{Y}^n) \propto \prod_{t=1}^{n} \prod_{j=1}^{J} \left[\frac{g_{j,t}(\boldsymbol{\beta})}{G_t(\boldsymbol{\beta})} \right]^{y_{j,t}}, \qquad (8.7)$$

where the subscript $\mathcal{M}$ stands for *multinomial*. To carry out Bayesian inference, we employ appropriate priors on $\boldsymbol{\beta}$ and obtain posterior summaries.

Templates for directly specifiying the formulas for Bayesian modeling of categorical data are not currently available in INLA. Instead, INLA uses the well-known *Multinomial-Poisson (M-P) transform* to rewrite the multinomial likelihood in (8.7) as an *equivalent Poisson likelihood* (see Baker (1994) and Guimaraes (2004)), an approach referred to in the literature as the *Poisson-trick* (Lee et al. (2017) and Chen and Liu (1997)).

Let ϕ_t be an auxiliary parameter for each time t, and let $\boldsymbol{\phi} = (\phi_1, \ldots, \phi_n)'$. The idea behind the Multinomial-Poisson (M-P) transformation is to assume that independently for all t, Y_{t+} is a random variable with

$$Y_{t+} \sim \text{Poisson}(\phi_t G_t(\boldsymbol{\beta})), \qquad (8.8)$$

and to show that estimating $\boldsymbol{\beta}$ by maximizing (8.7) is equivalent to estimating it by maximizing

$$L_{\mathcal{P}}(\boldsymbol{\phi}, \boldsymbol{\beta}; \boldsymbol{Y}^n) \propto \prod_{t=1}^{n} \prod_{j=1}^{J} [g_{j,t}(\boldsymbol{\beta})e^{\phi_t}]^{y_{j,t}} \exp[-g_{j,t}(\boldsymbol{\beta})e^{\phi_t}]. \qquad (8.9)$$

See Chapter 8 – Appendix for details.

8.4.1   Example: Simulated categorical time series

Consider a dynamic model for $\boldsymbol{y}_t = (Y_{1,t}, \ldots, Y_{J,t})'$ with

$$g_{j,t}(\boldsymbol{\beta}_t, \boldsymbol{Z}_{j,t}) = \exp(\boldsymbol{Z}'_{j,t}\boldsymbol{\beta}_t) = \exp(\eta_{j,t}), \text{where}$$

$$\eta_{j,t} = \begin{cases} \beta_{j,t,0} + \beta C_{j,t} + \delta_j U_{j,t} & \text{if } j = i, \\ \beta C_{j,t} + \delta_j U_{j,t} & \text{if } j \neq i, \end{cases} \qquad (8.10)$$

where the time-varying intercept $\beta_{j,t,0}$ for the selected category evolves as a random walk

$$\beta_{j,t,0} = \beta_{j,t-1,0} + w_{j,t}, \qquad (8.11)$$

with $w_{j,t} \sim N(0, \sigma^2_{w_j})$. For instance, if we let $j = 1$, then $\eta_{1,t}$ for the first category includes a dynamic intercept $\beta_{1,t,0}$, while the others do not. In (8.10), $C_{j,t}$ is a category specific predictor over time with a coefficient β which is constant across t and j, $U_{j,t}$ is another category specific predictor over time with a category-specific coefficient δ_j. For more on this setup, see.[1]

Model D1: Dynamic intercept in a single category

We generate a *single categorical time series* of length $n = 500$, with $J = 3$ categories, so that, at each time t, $N = 100$ units are categorized into these three groups. Here, N denotes the multinomial size and $Y_{j,t}$ denotes the multinomial counts for $j = 1, \ldots, J$ categories at time t. We set up the true parameter values for the model in (8.10) and (8.11), and generate a latent time-varying random effect $\beta_{1,t,0}$ for the *first* category.

```
beta <-  -0.3  # coefficient for C[j,t]
deltas <-  c(1, 4, 3) # category specific coefficients for U[j,t]
param <-  c(beta, deltas)
n <-  1000
set.seed(123457)
# category specific predictor with constant coefficient beta
C.1 <-  rnorm(n, mean = 30, sd = 2.5)
C.2 <-  rnorm(n, mean = 35, sd = 3.5)
C.3 <-  rnorm(n, mean = 40, sd = 1)
# category specific predictor with category-specific coefficient delta_j
U.1 <-  rnorm(n, mean = 2, sd = 0.5)
U.2 <-  rnorm(n, mean = 2, sd = 0.5)
U.3 <-  rnorm(n, mean = 2, sd = 0.5)
```

We set up the following function to simulate the time series from model (8.10) and (8.11). Note that we can include a random walk effect for either of the other categories $j = 2$ as $\beta_{2,t,0}$ or for $j = 3$ as $\beta_{3,t,0}$ instead of category $j = 1$, by modifying the code below, i.e., uncommenting `random.int[t]` for the category we wish to modify and commenting it out from the others.

```
Multinom.sample.rand <- function(N, random.int) {
  Y <- matrix(NA, ncol = 3, nrow = n)
  for (t in 1:n) {
    eta.1 <- beta * C.1[t] + deltas[1] * U.1[t] +
      random.int[t]
    eta.2 <- beta * C.2[t] + deltas[2] * U.2[t]
    #    + random.int[t]
    eta.3 <- beta * C.3[t] + deltas[3] * U.3[t]
    #    + random.int[t]
    probs <- c(eta.1, eta.2, eta.3)
    probs <- exp(probs) / sum(exp(probs))
    samp <- rmultinom(1, N, prob = probs)
    Y[t,] <- as.vector(samp)
```

---

[1]https://rdrr.io/GitHub/inbo/INLA/f/vignettes/multinomial.Rmd

```
  }
  colnames(Y) <- c("Y.1", "Y.2", "Y.3")
  return(Y)
}
```

We generate $\beta_{1,t,0}$ from a random walk with standard deviation 0.1.

```
rw1.int <- rep(NA, n)
rw1.int[1] <- 0
for (t in 2:n) {
  rw1.int[t] <- rnorm(1, mean = rw1.int[t - 1], sd = 0.1)
}
```

We then generate the multinomial time series $Y.rw1$ with size $N = 100$ at each time t, and plot the series, see Figure 8.4.

```
N <- 100
Y.rw1 <- Multinom.sample.rand(N, rw1.int)
```

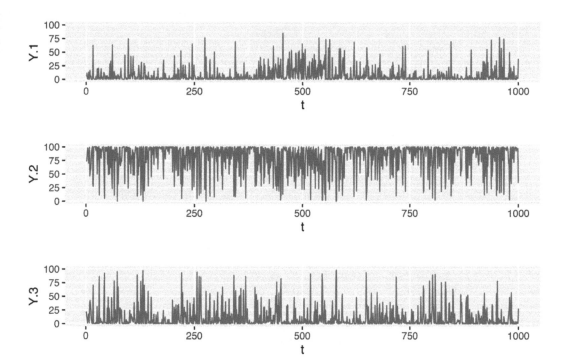

FIGURE 8.4: Simulated multinomial time series from Model D1.

Next, we build the data set to input into R-INLA. This data set will have $3n$ rows, one for each choice category in each choice situation, Y.1, Y.2, Y.3. For the predictor $C_{j,t}$, we use the long format, so we stack C.1, C.2, C.3 into a long vector C in a single column. For the predictors $U_{j,t}$, we use three columns, U.1, U.2, U.3, one for each category with $3n$ rows each. Each column will have the value of the predictor for the corresponding category, and zeros for the other categories.

The data frame additionally includes one more column, i.e., an index column representing the choice situation, with an additional parameter ϕ, corresponding to the ϕ_t we included in the M-P transformation. Note that ϕ_t indexes time but remains the same across categories for each time. The following function takes as input a data frame in wide format and transforms it into the structure described above, and sets up the index for the dynamic intercept $\beta_{1,t,0}$. We then construct the data frame *inla.dat*.

```
Y <- c(Y.rw1[,1], Y.rw1[, 2], Y.rw1[, 3])
C <- c(C.1, C.2, C.3)
U.1 <- c(U.1, rep(NA, 2*n))
U.2 <- c(rep(NA, n),U.2, rep(NA, n))
U.3 <- c(rep(NA, 2*n),U.3)
rw1.idx <- c(1:n, rep(NA, 2*n))
phi <- rep(1:n, 3)
inla.dat <- cbind.data.frame(Y, C, U.1, U.2, U.3, rw1.idx, phi)
```

The formula using the M-P transformation is given below.

```
formula.rw1 <- Y ~ -1 + C + U.1 + U.2 + U.3 +
  f(phi, model = "iid",
    hyper = list(prec = list(initial = -10,fixed = TRUE))) +
  f(rw1.idx, model = "rw1", constr = FALSE)
```

For ϕ_t, the default is f(phi, model = "iid", hyper = list(prec = list(initial = -10, fixed = TRUE))); but we can instead use f(phi, initial = -10, fixed = TRUE) and get the same results.

```
model.rw1 <- inla(
  formula.rw1,
  family = "poisson",
  data = inla.dat,
  control.predictor = list(compute = TRUE),
  control.compute = list(dic = TRUE, cpo = TRUE, waic = TRUE)
)
#summary(model.rw1)
format.inla.out(model.rw1$summary.fixed[, c(1:5)])
##   name mean    sd     0.025q  0.5q    0.975q
## 1 C      -0.298 0.003 -0.304 -0.298 -0.292
## 2 U.1    0.960 0.030  0.901  0.960  1.019
## 3 U.2    3.957 0.028  3.902  3.957  4.012
## 4 U.3    2.951 0.024  2.905  2.951  2.997
format.inla.out(model.rw1$summary.hyperpar[, c(1:2)])
##   name                    mean  sd
## 1 Precision for rw1.idx 119.3 24.96
```

The results show that the fixed effects and precision parameters from the simulation are recovered well. Figure 8.5 shows the simulated and fitted values of the random walk for $\beta_{1,t,0}$. We also plot the observed and fitted values for each category in Figure 8.6. We compute the in-sample MAE for each category, see Table 8.1.

Alternately, we can include the random intercept only in $\eta_{2,t}$ (Model D2) or only in $\eta_{3,t}$ (Model D3) instead of $\eta_{1,t}$. The in-sample MAEs for these situations are shown in Table 8.1. Detailed code can be seen in the GitHub link for the book.

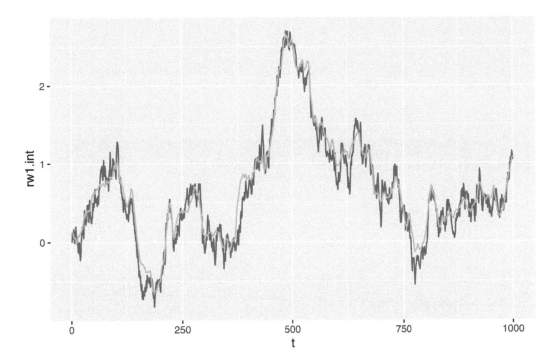

FIGURE 8.5: Actual values (red) and estimated (blue) random walk series for the dynamic intercept $\beta_{1,t,0}$ from Model D1.

Model D4: Dynamic coefficient for $C_{j,t}$

Consider the model with

$$\eta_{j,t} = \beta_t C_{j,t} + \delta_j U_{j,t}, \ j = 1, \ldots, 3 \qquad (8.12)$$

where the dynamic coefficient β_t evolves as a random walk

$$\beta_{t,0} = \beta_{t-1,0} + w_t, \qquad (8.13)$$

with $w_t \sim N(0, \sigma_w^2)$. We use the same true values of δ_j, $C_{j,t}$ and $U_{j,t}$ as before for the simulation. We source the function below for generating the multinomial responses corresponding to the model (8.12) and (8.13).

```
Multinom.sample.rand <- function(N, rw1.beta) {
  Y <- matrix(NA, ncol = 3, nrow = n)
  eta.1 <- eta.2 <- eta.3 <- rep(NA,n)
  for (t in 1:n) {
    eta.1[t] <- rw1.beta[t] * C.1[t] + deltas[1] * U.1[t]
    eta.2[t] <- rw1.beta[t] * C.2[t] + deltas[2] * U.2[t]
    eta.3[t] <- rw1.beta[t] * C.3[t] + deltas[3] * U.3[t]
    probs <- c(eta.1[t], eta.2[t], eta.3[t])
    probs <- exp(probs) / sum(exp(probs))
    samp <- rmultinom(1, N, prob = probs)
    Y[t,] <- as.vector(samp)
```

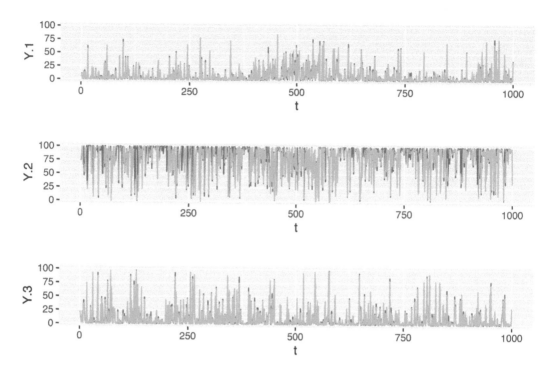

FIGURE 8.6: Actual values (red) and fitted (blue) responses from Model D1.

```
  }
  colnames(Y) <- c("Y.1", "Y.2", "Y.3")
  return(Y)
}
```

We first generate a random walk series corresponding to `rw1.beta`. We then compute and plot the response series (plots are not shown here).

```
rw1.beta <- rep(NA, n)
rw1.beta[1] <- 0
for (t in 2:n) {
  rw1.beta[t] <- rnorm(1, mean = rw1.beta[t - 1], sd = 0.1)
}
N <- 100
Y.rw1 <- Multinom.sample.rand(N, rw1.beta)
df.rw1 <- cbind.data.frame(Y.rw1, C.1, C.2, C.3, U.1, U.2, U.3)
```

We construct necessary indexes and the data frame as shown below.

```
Y.new <- c(Y.rw1[,1], Y.rw1[, 2], Y.rw1[, 3])
C.new <- c(df.rw1$C.1, df.rw1$C.2, df.rw1$C.3)
C.1.new <- c(C.1, rep(NA, 2*n))
C.2.new <- c(rep(NA, n), C.2, rep(NA, n))
C.3.new <- c(rep(NA, 2*n), C.3)
C.new <- c(C.1, C.2, C.3)
```

```
U.1.new <- c(U.1, rep(NA, 2*n))
U.2.new <- c(rep(NA, n), U.2, rep(NA, n))
U.3.new <- c(rep(NA, 2*n), U.3)
id.rw <- phi.new <- c(rep(1:n, 3))
rw1.idx <- c(1:n, rep(NA, 2*n))
rw2.idx <- c(rep(NA, n), 1:n, rep(NA, n))
rw3.idx <- c(rep(NA, 2*n), 1:n)
re.idx <- rep(1, (3*n))
id.re <- rep(1:n, 3)
inla.dat <- data.frame(Y.new, C.new, C.1.new, C.2.new, C.3.new,
                       U.1.new, U.2.new, U.3.new,
                       phi.new, id.rw, re.idx, id.re,
                       rw1.idx, rw2.idx, rw3.idx)
```

The formula and `inla()` call are shown below, followed by a summary of results.

```
formula.rw1 <- Y.new ~ -1 + U.1.new + U.2.new + U.3.new +
  f(phi.new, model = "iid",
    hyper = list(prec = list(initial = -10, fixed = TRUE))) +
  f(id.rw, C.new, model = "rw1", constr = FALSE) +
  f(
    id.re,
    model = "iid",
    replicate = re.idx,
    hyper = list(theta = list(initial = 20, fixed = T))
  )
model.rw1 <- inla(
  formula.rw1,
  family = "poisson",
  data = inla.dat,
  control.predictor = list(compute = TRUE),
  control.compute = list(dic = TRUE, cpo = TRUE, waic = TRUE)
)
#summary(model.rw1)

format.inla.out(model.rw1$summary.fixed[,c(1:5)])
##   name     mean  sd    0.025q 0.5q  0.975q
## 1 U.1.new 1.018 0.046 0.927  1.018 1.109
## 2 U.2.new 3.975 0.039 3.899  3.975 4.051
## 3 U.3.new 2.966 0.033 2.902  2.966 3.030
format.inla.out(model.rw1$summary.hyperpar[,c(1:2)])
##   name                    mean  sd
## 1 Precision for id.rw 116.9 10.01
```

The fixed and random coefficients are recovered well by the model fit. We can also plot the simulated and fitted values for the dynamic coefficient β_t as shown in Figure 8.7 below.

Plots of the observed and fitted responses are shown in Figure 8.8. In-sample MAEs from all four models, Model D1–D4, are shown in Table 8.1, suggesting that Model D4 performs the best.

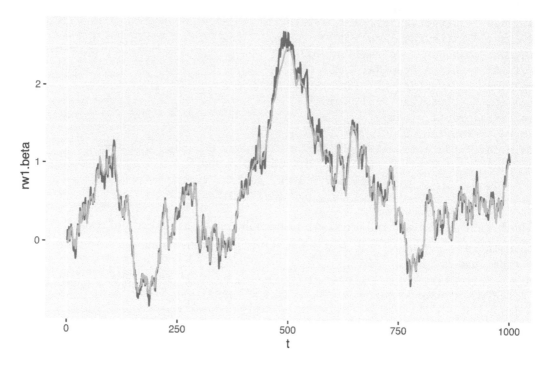

FIGURE 8.7: Actual values (red) and estimated (blue) random walk series for β_t from Model D4.

TABLE 8.1: In-sample MAE for model comparison.

Model	MAE(Y.1)	MAE(Y.2)	MAE(Y.3)
D1	1.362	2.058	1.551
D2	0.852	1.492	1.278
D3	1.034	1.969	1.665
D4	0.151	1.383	1.352

Chapter 8 – Appendix

Poisson-trick for multinomial models

Given the multinomial likelihood in (8.7), assume that Y_{t+} is a random variable with Poisson p.m.f. given in (8.8). For $t = 1, \ldots, n$, we can write

$$P(\boldsymbol{y}_t, Y_{t+} = y_{t+}) = P(Y_{t+} = y_{t+})P(\boldsymbol{y}_t | Y_{t+} = y_{t+}), \qquad (8.14)$$

where the first expression on the right side is given by (8.8) and the second expression, by (8.4). The marginal p.m.f. of $\boldsymbol{y}_t$ can be obtained by summing the joint p.m.f. in (8.14) over all possible values of Y_{t+} to yield the *Poisson* likelihood (which was given in (8.9)):

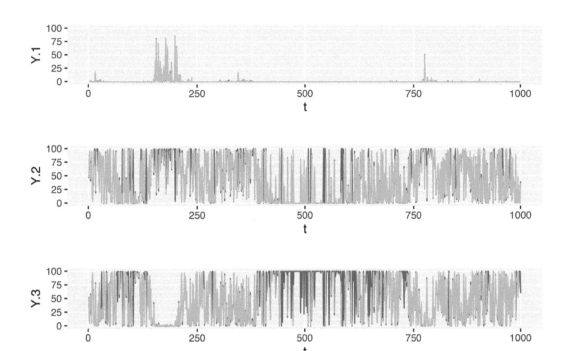

FIGURE 8.8: Actual values (red) and fitted (blue) responses from Model D4.

$$L_{\mathcal{P}}(\boldsymbol{\phi}, \boldsymbol{\beta}; \boldsymbol{Y}^n) \propto \prod_{t=1}^{n} \prod_{j=1}^{J} [g_{j,t}(\boldsymbol{\beta}) e^{\phi_t}]^{y_{j,t}} \exp[-g_{j,t}(\boldsymbol{\beta}) e^{\phi_t}].$$

Let

$$\widehat{\boldsymbol{\phi}}(\boldsymbol{\beta}) = \arg \max_{\boldsymbol{\phi}} L_{\mathcal{P}}(\boldsymbol{\phi}, \boldsymbol{\beta}; \boldsymbol{Y}^n) \tag{8.15}$$

denote the maximum of the profile likelihood function of $\boldsymbol{\phi}$ given $\boldsymbol{\beta}$. It can be shown that for $t = 1, \ldots, n$,

$$\widehat{\phi}_t(\boldsymbol{\beta}) = \frac{y_{t+}}{G_t(\boldsymbol{\beta})}. \tag{8.16}$$

Substituting $\widehat{\phi}_t(\boldsymbol{\beta})$ from (8.16) into (8.9), we can write (8.9) as a function of only $\boldsymbol{\beta}$, and show that the resulting expression $L_{\mathcal{P}}(\boldsymbol{\beta}; \boldsymbol{Y}^n)$ is equivalent to $L_{\mathcal{M}}(\boldsymbol{\beta}; \boldsymbol{Y}^n)$. It follows that estimation based on $L_{\mathcal{P}}(\boldsymbol{\phi}, \boldsymbol{\beta}; \boldsymbol{Y}^n)$ and $L_{\mathcal{M}}(\boldsymbol{\beta}; \boldsymbol{Y}^n)$ will be identical.

9

Modeling Count Time Series

9.1 Introduction

This chapter describes use of `R-INLA` for analyzing time series of counts. Section 9.2 describes models for a single univariate count time series. Section 9.3 shows how to fit a dynamic subject-specific model to a set of univariate count time series. To begin, we source the custom functions.

```
source("functions_custom.R")
```

This chapter uses the following packages.

```
library(INLA)
library(tidyverse)
```

We analyze the following data.

1. Crash counts in CT

2. Daily bike rentals

3. Ridesourcing in NYC

9.2 Univariate time series of counts

In the dynamic generalized linear model (DGLM) for a time series of counts, the response variable Y_t is assumed to have a suitable probability distribution which belongs to the exponential family, negative binomial or Poisson, say. The mean $E(Y_t) = \mu_t$ is linked to a structured predictor through a link function, $g(\mu_t) = \log(\mu_t)$. The structured predictor can take the form $\eta_t = \beta_{t,0} + \sum_{j=1}^{k} \beta_{t,j} Z_{t,j}$, where $Z_{t,j}$, $j = 1, \ldots, k$ are exogenous predictors, and the coefficients $\beta_{t,j}$, $j = 0, \ldots, k$ are regarded as the state parameters in the DGLM, and may evolve randomly as latent Gaussian processes depending on some hyperparameters.

9.2.1 Example: Simulated univariate Poisson counts

We simulate a Poisson time series with mean μ_t. The observation model and state equation of the DGLM are

$$Y_t|\mu_t \sim \text{Poisson}(\mu_t),$$
$$\log(\mu_t) = \beta_t,$$
$$\beta_t = \phi\beta_{t-1} + w_t, \tag{9.1}$$

where, β_t is an unknown state variable, $|\phi| < 1$, and w_t is $N(0, \sigma_w^2)$. The count time series corresponding to true values of $\phi = 0.65$ and $\sigma_w^2 = 0.25$ is shown in Figure 9.1.

```
sim.y.ar1.pois <-
  simulation.from.ar.poisson(
    sample.size = 500,
    burn.in = 100,
    phi = 0.65,
    level = 0,
    drift = 0,
    W = 0.25,
    plot.data = TRUE,
    seed = 123457
  )
sim.y.ar1.pois$sim.plot
```

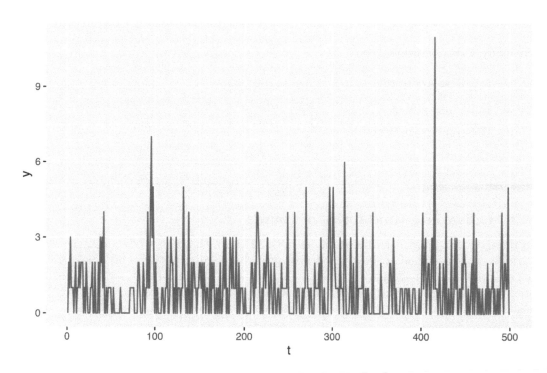

FIGURE 9.1: Time series of simulated Poisson counts.

```
y.ar1.pois <- sim.y.ar1.pois$sim.data
n <- length(sim.y.ar1.pois$sim.data)
```

The formula for fitting (9.1) is shown below. Note that in the `inla()` function, we set the argument `control.predictor = list(compute = TRUE, link = 1)`. The argument compute = TRUE allows the marginals of the linear predictor to be computed. The argument link = 1 specifies the log link corresponding to the Poisson family. Note that, if we do not specify link=1, the identity link would be used by INLA for the hold-out fits, i.e., η_t will be linked to μ_t instead of $\log(\mu_t)$. The results from the posterior distributions of the hyperparameters ϕ and σ_w^2 show that we are able to recover the true values in the simulation.

```
id <- 1:n
sim.pois.dat <- cbind.data.frame(y.ar1.pois, id)
formula <- y.ar1.pois ~ f(id, model = "ar1") - 1
model.sim.pois <- inla(
  formula,
  family = "poisson",
  data = sim.pois.dat,
  control.predictor = list(compute = TRUE, link = 1)
)
#summary(model.sim.pois)
format.inla.out(model.sim.pois$summary.hyperpar[,c(1:2)])
##   name              mean   sd
## 1 Precision for id  3.137  0.755
## 2 Rho for id        0.590  0.124
```

We first use inla.tmarginal to obtain the marginal distribution of the state variance, and then compute its 95% HPD interval using inla.hpdmarginal() as shown below.

```
state.var <-  inla.tmarginal(
  fun = function(x)
    exp(-x),
  marginal =
    model.sim.pois$internal.marginals.hyperpar$`Log precision for id`
)
var.state.hpd.interval <- inla.hpdmarginal(p = 0.95, state.var)
round(var.state.hpd.interval, 3)
##              low    high
## level:0.95 0.193  0.492
```

In the code below, summary.fitted.values returns summaries of the posterior marginals of the fitted values which are obtained by transforming the linear predictors by the inverse of the link function.

```
head(model.sim.pois$summary.fitted.values$mean) %>%
  round(3)
## [1] 1.012 1.293 1.318 1.582 1.255 1.123
```

However, note that these cannot be obtained by directly applying the inverse transformation to summary.linear.predictor, which returns summaries of posterior marginals of the linear predictors in the model.. For example, under a log link, we cannot just apply the exp() function to summary.linear.predictor$mean, since this will not result in summary.fitted.values$mean.

```
exp(head(model.sim.pois$summary.linear.predictor$mean)) %>%
  round(3)
## [1] 0.906 1.173 1.200 1.444 1.142 1.018
```

We must instead transform the marginals for the variable *fitted.from.lp*.

```
fitted.from.lp <- model.sim.pois$marginals.linear.predictor %>%
  sapply(function(x)
    inla.emarginal(
      fun = function(y)
        exp(y),
      marginal = x
    ))
head(fitted.from.lp) %>%
  round(3)
## Predictor.1 Predictor.2 Predictor.3 Predictor.4 Predictor.5
##        1.012       1.293       1.318       1.582       1.255
## Predictor.6
##        1.122
```

Fitted values, i.e., means of the posterior predictive distributions (blue), along with the interval between the 2.5^{th} and 97.5^{th} percentiles (gray shaded region), are shown in Figure 9.2.

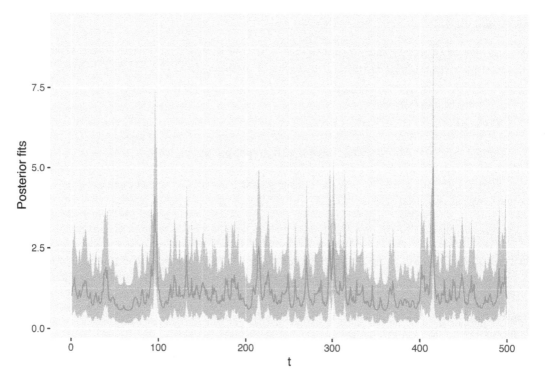

FIGURE 9.2: Means of the posterior predictive distributions (blue) and the interval between the 2.5^{th} and 97.5^{th} percentiles (gray shaded region).

9.2.2   Example: Modeling crash counts in CT

Hu et al. (2013) described a DGLM in the context of transportation safety. The data[1] is obtained from records for all crashes on state-maintained roads in Connecticut from January

---

[1]https://www.ctcrash.uconn.edu/

1995 to December 2009 collected from the Connecticut Department of Transportation (ConnDOT). Crashes are grouped by level of severity, which include fatal (K), severe injury (A), evident minor injury (B), non-evident possible injury (C), and property damage only (O). For each month, crash counts are aggregated at each severity level by access type (surface or limited access roads), area type (rural or urban), and whether or not the driver of at least one of the vehicles involved in the crash was a senior. Severity for a crash is defined as the highest injury severity experienced by any individual involved. Monthly average daily vehicle miles traveled (VMT) in each of these four categories are obtained by applying monthly expansion factors (obtained from ConnDOT) to annual average daily VMTs (see Hu et al. (2013) for details). We model a univariate time series of monthly KAB crash counts, by cumulating crashes of type K, A, or B that involved senior drivers and occurred on rural, limited access roads. The predictor is the VMT, denoted here by V_t. We use a negative binomial DGLM with $\mu_t = E(Y_t)$, and shape parameter κ. The observation model of the DGLM is

$$Y_t|\mu_t \sim \text{NegBin}(\mu_t, k), \text{ where,}$$
$$\mu_t = \alpha_t V_t^{\beta_{t,1}}, \text{ i.e.,}$$
$$\log(\mu_t) = \beta_{t,0} + \beta_{t,1} Z_t, \tag{9.2}$$

where $\beta_{t,0} = \log(\alpha_t)$ and $Z_t = \log(V_t)$. Let $\boldsymbol{\beta}_t = (\beta_{t,0}, \beta_{t,1})'$ be a 2-dimensional state (parameter) vector, which evolves according to the state equations shown below:

$$\beta_{t,0} = \beta_{t-1,0} + w_{t,0},$$
$$\beta_{t,1} = \beta_{t-1,1} + w_{t,1}, \tag{9.3}$$

where $w_{t,0} \sim N(0, W_0)$ and $w_{t,1} \sim N(0, W_1)$. The time series is shown in Figure 9.3.

```
dat <-
  read.csv(file = "crashcount_rle.csv",
           header = TRUE,
           stringsAsFactors = FALSE)
n <- nrow(dat)
kabct <- dat$KAB_FREQ
vmt <-  log(dat$vmt_exp)
tsline(kabct,
       xlab = "t",
       ylab = "crash counts",
       line.color = "red", line.size = 0.6)
```

The indexes, data frame, and formula are as follows.

```
id.b0 <- id.b1 <- 1:n
data.crash <- data.frame(id.b0, id.b1, kabct, vmt)
formula.crashct <-
  kabct ~  -1 + f(id.b0,
              model = "rw1",
              initial = 5,
              constr = FALSE) +
  f(id.b1,
    vmt,
    model = "rw1",
```

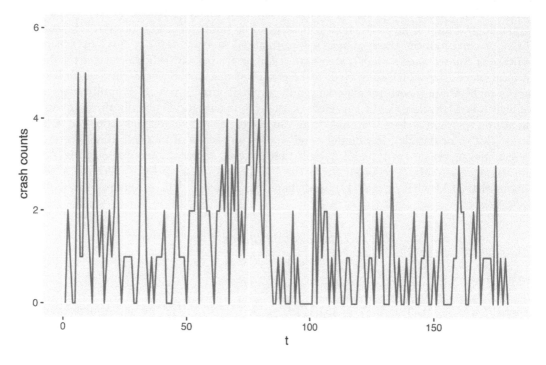

FIGURE 9.3: Time series of crash counts.

```
    initial = 5,
    constr = FALSE)
```

A summary of the results is shown below. The plots in Figure 9.4 show the posterior means of the state variables, i.e., the dynamic intercept on the log scale ($\beta_{t,0}$), and the dynamic slope $\beta_{t,1}$.

```
model.crashct = inla(
  formula.crashct,
  family = "nbinomial",
  data = data.crash,
  control.compute = list(
    dic = TRUE,
    waic = TRUE,
    cpo = TRUE,
    config = TRUE
  ),
  control.family = list(hyper = list(theta = list(
    prior = "loggamma",
    param = c(1, 1)
  ))))
)
#summary(model.crashct)
format.inla.out(model.crashct$summary.hyperpar[, c(1:2)])
##    name                                        mean       sd
```

```
## 1 size for nbinomial obs 1/overdispersion       2.942      0.974
## 2 Precision for id.b0                      19671.628 21737.167
## 3 Precision for id.b1                      13947.980 41998.126
```

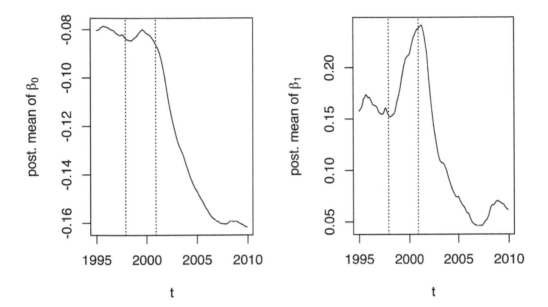

FIGURE 9.4: Posterior means of intercepts on the log scale (left) and the slope (right).

The vertical dotted lines in Figure 9.4 correspond to September 1997 and September 2000, when stricter passenger car regulation standards were implemented for airbags and anti-lock brakes, respectively. It is interesting to see how the estimated posterior means of $\beta_{t,0}$ and $\beta_{t,1}$ behave over time, and how they respond to the two events. We see that on rural limited access roads, senior drivers show a decreasing trend in the posterior mean of $\beta_{t,0}$ after the second intervention, and decreasing pattern in the posterior mean of $\beta_{t,1}$, with a bump at the end. Overall, there are interesting temporal patterns for the KAB injury crashes on these selected roads.

A *static* negative binomial loglinear model (code shown below) can be compared with the DGLM fit; we do not show these results here, but they are available in the GitHub link for the book.

```
reg.negbin <- MASS::glm.nb(kabct ~ vmt, link = log)
summary(reg.negbin)
```

9.2.3   Example: Daily bike rentals in Washington D.C.

We use data on daily bike sharing customers and weather in the metro DC area from January 1, 2011 to December 31, 2018. The data is obtained from Kaggle.[2]

---

[2]https://www.kaggle.com/juliajemals/bike-sharing-washington-dc.

We model the daily bike usage as a function of temperature (Fahrenheit), humidity, and wind speed. We set aside the last $n_h = 7$ data points for validating the model fit.

```
bike <- read.csv("bikeusage_daily.csv")
n <- nrow(bike)
n.hold <- 7
n.train <- n - n.hold
test.data <- tail(bike$Total.Users, n.hold)
train.data <- bike[1:n.train, ]
```

The time series (see Figure 9.5) exhibits periodic behavior, with period $s = 7$. We set up a model with trend and seasonality (with period $s = 7$), and the weather variables as predictors.

```
n.seas <- s <-  7
y <- train.data$Total.Users
y.append <- c(y, rep(NA, n.hold))
id <- id.trend <-  id.seas <- 1:length(y.append)
bike.dat <-
  cbind.data.frame(id,
                   id.trend,
                   id.seas,
                   y.append,
                   select(bike, Temperature.F:Wind.Speed))
```

We fit a Poisson DGLM, treating only the trend and seasonal effects as random (similar to the structural time series model in Chapter 5), while the weather effects are treated as fixed effects. The observation model is

$$Y_t|\mu_t \sim \text{Poisson}(\mu_t),$$
$$\log(\mu_t) = m_t + S_t + \boldsymbol{z}_t'\boldsymbol{\beta}, \tag{9.4}$$

where m_t and S_t are the random trend and seasonal effects modeled as shown below, and $\boldsymbol{\beta}$ is a vector of static coefficients corresponding to the exogenous predictors $\boldsymbol{z}_t$. We model the latent state variables by

$$m_t = m_{t-1} + w_{t,1},$$
$$S_t = -\sum_{j=1}^{6} S_{t-j} + w_{t,2}, \tag{9.5}$$

where $w_{t,j} \sim N(0, \sigma_{w_j}^2)$, $j = 1, 2$. The formula and fit are shown below.

```
bike.formula <- y.append ~  Temperature.F + Humidity + Wind.Speed +
  f(id.trend, model = "rw1") +
  f(id.seas, model = "seasonal", season.length = n.seas)
bike.model <- inla(
  bike.formula,
  family = "poisson",
  data = bike.dat,
  control.predictor = list(compute = TRUE, link = 1),
  control.compute = list(
    dic = TRUE,
```

```
      waic = TRUE,
      cpo = TRUE,
      config = TRUE
   ),
   control.family = list(link = "log")
)
#summary(bike.model)
format.inla.out(bike.model$summary.fixed[,c(1:5)])
##    name              mean    sd     0.025q  0.5q    0.975q
## 1 Intercept          5.035 0.129   4.782   5.035   5.287
## 2 Temperature.F      0.015 0.002   0.012   0.015   0.019
## 3 Humidity          -0.011 0.001  -0.012  -0.011  -0.009
## 4 Wind.Speed        -0.004 0.001  -0.007  -0.004  -0.002
format.inla.out(bike.model$summary.hyperpar[,c(1:2)])
##    name                          mean   sd
## 1 Precision for id.trend 24.33 4.253
## 2 Precision for id.seas  44.73 8.381
```

Using the approach described in Section 9.2.1, we obtain fitted values using `summary.fitted.values$mean`. These are plotted in Figure 9.5.

```
fit.bike <- bike.model$summary.fitted.values$mean
fit.bike.df <-
  cbind.data.frame(
    time = 1:n,
    obs.bike = bike$Total.Users,
    fit.bike = fit.bike
  )
multiline.plot(
  fit.bike.df,
  line.size = 0.5,
  xlab = "t",
  ylab = "bike counts",
  line.color = c("red", "blue")
)+
  theme(legend.position = "none")
```

We compute out-of-sample forecasts of the hold-out data, and then compute the MAE.

```
yfore <- tail(fit.bike, n.hold)
mae(test.data, round(yfore))
## [1] "MAE is 50.143"
```

Note that we can fit other models to this data by (a) selecting the best subset of exogenous predictors, (b) replacing the random walk evolution for m_t by other processes, or (c) replacing the Poisson sampling distribution for Y_t by the negative binomial distribution. We can compare the different models for bike usage using criteria such as the DIC, WAIC, and MAE.

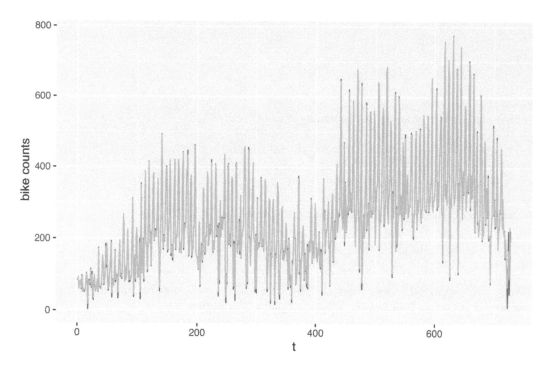

FIGURE 9.5: Observed (red) and fitted (blue) bike counts.

9.3   Hierarchical modeling of univariate count time series

This section describes fitting a hierarchical (subject-specific) dynamic model to a set of g
time series of counts, each of length n. In Section 9.3.1, we present the code and results for
three scenarios used to model several univariate time series of counts. In Section 9.3.2, we
illustrate such models for *daily* TNC counts over three years for taxi zones in Manhattan
(see Section 6.3).

9.3.1   Example: Simulated univariate Poisson counts

Let y_{it} denote the counts at time t from the ith series, for $t = 1, \ldots, n$ and $i = 1, \ldots, g$. Let
$E(y_{it}) = \lambda_{it}$. For simplicity, we first assume that there are no exogenous predictors, and
build the following model:

$$
\begin{aligned}
y_{it}|\lambda_{it} &\sim \text{UDC}(\lambda_{it}), \\
\log \lambda_{it} &= \alpha_i + \beta_{it,0}, \\
\beta_{it,0} &= \phi\beta_{i,t-1,0} + w_{it},
\end{aligned}
\tag{9.6}
$$

where, we assume that UDC refers to *univariate distribution of counts*, such as the Poisson,
negative binomial, or zero-inflated Poisson (ZIP) distributions. We allow the intercept $\beta_{it,0}$
to be random, varying over i and t, evolving over time as an AR(1) process. We assume that

w_{it} are $N(0, \sigma_w^2)$ variables, and UDC refers to the Poisson distribution. We discuss three separate cases, based on our assumption about the level.

Case 1. Common level model

We allow the $g = 300$ series to have the same level α, i.e., $\alpha_i = \alpha$ in (9.6). In the simulation, we allow the AR coefficient ϕ to vary slightly around a true value of 0.8 across the g series, and let $\alpha = 1.3$.

```
set.seed(123457)
g <- 300
n <- 100
burn.in <- 500
n.tot <- n + burn.in
sigma2.w <- 0.25
phi.0 <- 0.8
alpha <- 1.3
yit.all <- yit.all.list <- z1.all <- z2.all <- vector("list", g)
for (j in 1:g) {
  yit <- beta.0t <- lambda.t <- eta.t <- vector("double", n.tot)
  u <- runif(1, min = -0.1, max = 0.05)
  phi <- phi.0 + u
  beta.0t <-
    arima.sim(list(ar = phi, ma = 0), n = n.tot, sd = sqrt(sigma2.w))
  for (i in 1:n.tot) {
    eta.t[i] <- alpha + beta.0t[i]
    lambda.t[i] <- exp(eta.t[i])
    yit[i] <- rpois(1, lambda.t[i])
  }
  yit.all.list[[j]] <- tail(yit, n)
}
yit.all <- unlist(yit.all.list)
```

We set up indexes, data frame, and formula, and fit the common level model.

```
id.beta0 <- rep(1:n, g)
re.beta0 <- rep(1:g, each = n)
data.hpois.alpha.1 <-
  cbind.data.frame(id.beta0, yit.all, re.beta0)
formula.hpois.alpha.1 <-
  yit.all ~ 1 + f(id.beta0, model = "ar1", replicate = re.beta0)

model.hpois.alpha.1 <- inla(
  formula.hpois.alpha.1,
  family = "poisson",
  data = data.hpois.alpha.1,
  control.predictor = list(compute = TRUE, link = 1),
  control.compute = list(
    dic = TRUE,
    cpo = TRUE,
    waic = TRUE,
    config = TRUE,
    cpo = TRUE))
```

The results for the fixed and random parameters are shown below. We see that all the true parameters in the simulation are adequately recovered.

```
format.inla.out(summary.hpois.alpha.1$fixed)[c(1:5)]
##    name        mean  sd    0.025q 0.5q  0.975q mode   kld
## 1 Intercept 1.294 0.013 1.268  1.294 1.32   1.294 0
##         name  mean    sd 0.025q  0.5q
## 1 Intercept 1.294 0.013  1.268 1.294
format.inla.out(summary.hpois.alpha.1$hyperpar[c(1,2,4,5)])
##    name                        mean  sd    0.5q  0.975q
## 1 Precision for id.beta0 1.581 0.031 1.583 1.638
## 2 Rho for id.beta0         0.780 0.005 0.780 0.789
```

Alternatively, we can fit the model under the following non-default prior specification,

```
prior.hpois <- list(
    prec = list(prior = "loggamma", param = c(0.1, 0.1)),
    rho = list(prior = "normal", param = c(0, 0.15))
)
```

and then modify the following line of code in the `formula` as

```
formula.hpois.alpha.1.nd.prior <-
  yit.all ~ 1 + f(id.beta0,
                  model = "ar1",
                  replicate = re.beta0,
                  hyper = prior.hpois)
```

Results are available on our GitHub link.

Case 2. Model with different fixed levels

We simulate data from the model (9.6) with different fixed levels α_i, $i = 1, \ldots, g$, for $g = 10$. We allow the coefficient of an exogenous predictor Z_{it} to evolve dynamically as an AR(1) model with coefficient $\phi = 0.8$ and $\sigma_w^2 = 0.25$.

```
set.seed(123457)
g <- 10
n <- 100
burn.in <- 500
n.tot <- n + burn.in
sigma2.w0 <- 0.25
phi.0 <- 0.8
```

We slightly perturb the values of ϕ and σ_w^2 to reflect variability that would occur in practice. We use `arima.sim()` to generate underlying AR(1) models for the latent states and build the responses and predictors.

```
yit.all <-
  yit.all.list <-
  z1.all <- z2.all <- alpha <- z.all <- sigma2.w <- vector("list", g)
for (j in 1:g) {
  yit <- beta.1t <- lambda.t <- eta.t <- z <- vector("double", n.tot)
  u <- runif(1, min = -0.1, max = 0.05)
  phi <- phi.0 + u
```

```
us <- runif(1, min = -0.15, max = 0.25)
sigma2.w[[j]] <- sigma2.w0 + us
beta.1t <-
  arima.sim(list(ar = phi, ma = 0), n = n.tot, sd = sqrt(sigma2.w[[j]]))
alpha[[j]] <- sample(seq(-2, 2, by = 0.1), 1)
z <- rnorm(n.tot)
for (i in 1:n.tot) {
  eta.t[i] <- alpha[[j]] + beta.1t[i]  * z[i]
  lambda.t[i] <- exp(eta.t[i])
  yit[i] <- rpois(1, lambda.t[i])
}
yit.all.list[[j]] <- tail(yit, n)
z.all[[j]] <- tail(z, n)
}
yit.all <- unlist(yit.all.list)
z.all <- unlist(z.all)
alpha.all <- unlist(alpha)
```

The indexes, replicates, and formula are set up as shown below. The fixed α_i's are handled by `fact.alpha <- as.factor(id.alpha)`.

```
id.beta1 <- rep(1:n, g)
re.beta1 <- id.alpha <- rep(1:g, each = n)
fact.alpha <- as.factor(id.alpha)
formula.hpois.alpha.2 <-
  yit.all ~ -1 + fact.alpha +
  f(id.beta1, z.all, model = "ar1", replicate = re.beta1)
data.hpois.alpha.2 <-
  cbind.data.frame(id.beta1, z.all, yit.all, re.beta1, fact.alpha)
```

The `inla()` call for `model.hpois.alpha.2` is similar to the code for `model.hpois.alpha.1`, except for the replaced inputs `formula.hpois.alpha.2` and `data.hpois.alpha.2`. They produce the results shown below.

```
format.inla.out(summary.hpois.alpha.2$fixed)[c(1:5)]
##     name          mean   sd     0.025q 0.5q    0.975q mode    kld
## 1   fact.alpha1   -2.530 0.328  -3.225 -2.513  -1.933 -2.478  0
## 2   fact.alpha2   -1.882 0.250  -2.403 -1.872  -1.421 -1.851  0
## 3   fact.alpha3   -1.917 0.248  -2.432 -1.907  -1.458 -1.887  0
## 4   fact.alpha4   -0.613 0.145  -0.908 -0.610  -0.338 -0.603  0
## 5   fact.alpha5   -1.182 0.180  -1.551 -1.177  -0.845 -1.165  0
## 6   fact.alpha6    1.425 0.060   1.306  1.426   1.541  1.427  0
## 7   fact.alpha7   -1.025 0.170  -1.374 -1.020  -0.706 -1.010  0
## 8   fact.alpha8    0.043 0.111  -0.181  0.045   0.255  0.050  0
## 9   fact.alpha9    1.285 0.063   1.160  1.286   1.406  1.287  0
## 10  fact.alpha10  -0.693 0.150  -0.998 -0.689  -0.410 -0.682  0
##            name      mean    sd 0.025q    0.5q
## 1   fact.alpha1 -2.530 0.328  -3.225  -2.513
## 2   fact.alpha2 -1.882 0.250  -2.403  -1.872
## 3   fact.alpha3 -1.917 0.248  -2.432  -1.907
## 4   fact.alpha4 -0.613 0.145  -0.908  -0.610
## 5   fact.alpha5 -1.182 0.180  -1.551  -1.177
```

```
## 6    fact.alpha6   1.425 0.060  1.306  1.426
## 7    fact.alpha7  -1.025 0.170 -1.374 -1.020
## 8    fact.alpha8   0.043 0.111 -0.181  0.045
## 9    fact.alpha9   1.285 0.063  1.160  1.286
## 10 fact.alpha10  -0.693 0.150 -0.998 -0.689
format.inla.out(summary.hpois.alpha.2$hyperpar[c(1,2,4,5)])
##    name                    mean   sd     0.5q   0.975q
## 1 Precision for id.beta1 1.836 0.296 1.813 2.482
## 2 Rho for id.beta1        0.824 0.043 0.828 0.896
```

The estimated α_i's recover the true levels shown below:

```
alpha.all
##   [1] -1.8 -1.6 -2.0 -0.8 -1.1   1.4 -1.2   0.0   1.3 -0.9
```

Case 3. Model with different random levels

Let α_i's be independent $N(0, \sigma_\alpha^2)$ random variables, an assumption which is particularly useful when g is large, and we are only interested in investigating whether there is significant variation in the large population of time series from which g series are randomly drawn. The true parameters are the similar to Case 2, but $g = 500$ instead of 10. We set `alpha[[j]]` `<- sample(seq(-2, 2, by = 0.1), 1)` and generate the responses and predictors.

```
yit.all <-
  yit.all.list <-
  z1.all <- z2.all <- alpha <- z.all <- sigma2.w <- vector("list", g)
for (j in 1:g) {
  yit <- beta.1t <- lambda.t <- eta.t <- z <- vector("double", n.tot)
  u <- runif(1, min = -0.1, max = 0.05)
  phi <- phi.0 + u
  us <- runif(1, min = -0.15, max = 0.25)
  sigma2.w[[j]] <- sigma2.w0 + us
  beta.1t <-
    arima.sim(list(ar = phi, ma = 0), n = n.tot, sd = sqrt(sigma2.w[[j]]))
  alpha[[j]] <- sample(seq(-2, 2, by = 0.1), 1)
  z <- rnorm(n.tot)
  for (i in 1:n.tot) {
    eta.t[i] <- alpha[[j]] + beta.1t[i]   * z[i]
    lambda.t[i] <- exp(eta.t[i])
    yit[i] <- rpois(1, lambda.t[i])
  }
  yit.all.list[[j]] <- tail(yit, n)
  z.all[[j]] <- tail(z, n)
}
yit.all <- unlist(yit.all.list)
z.all <- unlist(z.all)
```

Unlike Case 2, we set `id.alpha <- rep(1:g, each = n)` and include it as a random effect in the formula. The `inla()` call is similar to Case 2 except for replacing that formula and data frame by `formula.hpois.alpha.3` and `data.hpois.alpha.3`. The results are shown below.

```
id.beta1 <- rep(1:n, g)
```

```
re.beta1 <- id.alpha <- rep(1:g, each = n)
prec.prior <- list(prec = list(param = c(0.001, 0.001)))
formula.hpois.alpha.3 <-
  yit.all ~ -1 + f(id.alpha, model="iid") +
  f(id.beta1, z.all, model = "ar1", replicate = re.beta1)
data.hpois.alpha.3 <-
  cbind.data.frame(id.beta1, z.all, yit.all, re.beta1, id.alpha)

format.inla.out(summary.hpois.alpha.3$hyperpar[c(1,2,4,5)])
##   name                   mean    sd     0.5q   0.975q
## 1 Precision for id.alpha 0.703  0.064  0.693  0.853
## 2 Precision for id.beta1 1.311  0.026  1.309  1.368
## 3 Rho for id.beta1              0.780  0.005  0.780  0.790
```

Similar to Case 2, we are able to recover the true ϕ and σ^2. The estimated posterior precision matches the true precision of the simulated random α's. The estimation accuracy will increase as g increases.

9.3.2 Example: Modeling daily TNC usage in NYC

We fit hierarchical count data models to *daily* TNC usage in the borough of Manhattan in NYC, consisting of 36 taxi zones. The variables are similar to what we have seen in Chapter 6 for weekly ridesourcing data. In almost every taxi zone in Manhattan, TNC usage shows an upward trend, Taxi usage shows a slight negative trend, while Subway usage seems relatively flat. Further, both TNC and Taxi usage exhibit periodicity, with seasonal period $s = 7$.

```
ride.3yr <-
  read.csv("tnc_daily_data.csv",
           header = TRUE,
           stringsAsFactors = FALSE)
ride <- ride.3yr
ride <- ride %>%
  filter(borough == "Manhattan")
n <- n_distinct(ride$date)
g <- n_distinct(ride$zoneid)
uid <- unique(ride$zoneid) #36 zones
ride.zone <- ride
```

Figure 9.6 shows scaled TNC and Taxi usage for the first nine zones.

To understand the behavior of daily TNC counts over the three years, we first investigate separate DLM fits to the data from each of the 36 zones. For simplicity, we have suppressed the subscript i in the *individual zone* models below. In a given zone, we let y_t denote the daily TNC usage on day t with $E(y_t) = \lambda_t$, and let UDC denote the Poisson distribution. We investigate three different *separate zone* models, Models Z1–Z3, which capture the time evolution in different ways.

Model Z1

Model Z1 includes a fixed level α, a fixed linear trend γt, a random dynamic intercept $\beta_{t,0}$ which is modeled as a random walk process, effects of predictors $\cos(2\pi t/7)$ and $\sin(2\pi t/7)$ to handle the seasonality, and effects of Taxi and Subway usage as fixed predictors. We denote these fixed predictors by $Z_{t,j}$, $j = 1, \ldots, 4$.

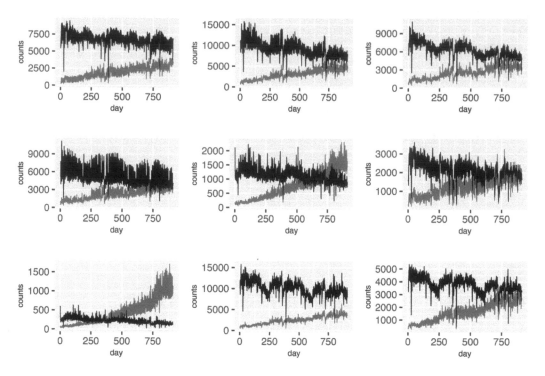

FIGURE 9.6: Time series plots of daily TNC (red) and Taxi (black) counts in nine different taxi zones in Manhattan.

$$y_t | \lambda_t \sim \text{Poisson}(\lambda_t),$$

$$\log \lambda_t = \alpha + \beta_{t,0} + \gamma t + \sum_{j=1}^{4} \beta_j Z_{t,j}$$

$$\beta_{t,0} = \beta_{t-1,0} + w_{t,0}, \qquad\qquad (9.7)$$

where $w_{t,0}$ is $N(0, \sigma_{w0}^2)$. We set up the indexes and arrays for collecting output from the different zones.

```
id.t <- 1:n
trend <- 1:n
seas.per <- 7
costerm <- cos(2*pi*trend/seas.per)
sinterm <- sin(2*pi*trend/seas.per)
corr.zone <- vector("list", g)
prec.int <-  vector("list", g)
alpha.coef  <-  taxi.coef <- subway.coef <- vector("list", g)
trend.coef <- cos.coef <- sin.coef <- vector("list", g)
dic.zone <- waic.zone <- mae.zone <- vector("list",g)
```

We next loop across the g zones, fitting Model Z1 in each zone, and saving the results. Note that we also investigate the correlation between the stationary time series of model residuals for TNC and Taxi after detrending and deseasonalizing both series.

```
for (k in 1:g) {
  ride.temp <- ride.zone[ride.zone$zoneid == uid[k], ]
  ride.temp$tnc <- round(ride.temp$tnc / 100)
  ride.temp$taxi <- round(ride.temp$taxi / 100)
  ride.temp$subway <- round(ride.temp$subway / 1000)
  zone.dat <-
    cbind.data.frame(ride.temp, trend, costerm, sinterm, id.t)
  lm.fit.tnc <-
    lm(ride.temp$tnc ~ trend + costerm + sinterm, data = zone.dat)
  res.tnc <- lm.fit.tnc$resid
  lm.fit.taxi <- lm(ride.temp$taxi ~ trend + sin(2 * pi * trend / 7) +
                    cos(2 * pi * trend / 7),
                  data = zone.dat)
  res.taxi <- lm.fit.taxi$resid
  corr.zone[[k]] <- cor(res.tnc, res.taxi)
  #ccf(res.tnc, res.taxi)
  formula.zone.Z1 <-
    tnc ~ 1 + trend + costerm + sinterm + taxi + subway +
    f(id.t, model = "rw1", constr = FALSE)
  model.zone.Z1 <- inla(
    formula.zone.Z1,
    family = "poisson",
    data = zone.dat,
    control.predictor = list(compute = TRUE, link = 1),
    control.compute = list(
      dic = TRUE,
      cpo = TRUE,
      waic = TRUE,
      config = TRUE,
      cpo = TRUE
    )
  )
  prec.int[[k]] <- model.zone.Z1$summary.hyperpar$mean[1]
  alpha.coef[[k]] <- model.zone.Z1$summary.fixed$mean[1]
  trend.coef[[k]] <- model.zone.Z1$summary.fixed$mean[2]
  cos.coef[[k]] <- model.zone.Z1$summary.fixed$mean[3]
  sin.coef[[k]] <- model.zone.Z1$summary.fixed$mean[4]
  taxi.coef[[k]] <- model.zone.Z1$summary.fixed$mean[5]
  subway.coef[[k]] <- model.zone.Z1$summary.fixed$mean[6]
  dic.zone[[k]] <- model.zone.Z1$dic$dic
  waic.zone[[k]] <- model.zone.Z1$waic$waic
  fit.zone <- model.zone.Z1$summary.fitted.values$mean
  mae.zone[[k]] <- sum(abs(ride.temp$tnc - fit.zone)) / n
}
```

To see the results, we unlist the saved output from all g zones. There is moderate contemporaneous correlation between TNC and Taxi usage (after detrending and deseasonalizing to make each series more or less stationary); see Figure 9.7. The `ccf()` plots (not shown) do not show much evidence that Taxi usage dynamically drives TNC usage. This is reflected in Models Z1–Z3 by not including a dynamic evolution for its coefficient, β_3.

```
model.zone.Z1.result <-
  list(
    corr.zone = unlist(corr.zone),
    prec.int = unlist(prec.int),
    alpha.coef = unlist(alpha.coef),
    trend.coef = unlist(trend.coef),
    cos.coef = unlist(cos.coef),
    sin.coef = unlist(sin.coef),
    taxi.coef = unlist(taxi.coef),
    subway.coef = unlist(subway.coef),
    dic.zone = unlist(dic.zone),
    waic.zone = unlist(waic.zone),
    mae.zone = unlist(mae.zone)
  )

out.corr <- model.zone.Z1.result$corr.zone
summary(out.corr)
##    Min. 1st Qu.  Median    Mean 3rd Qu.    Max.
##   0.111   0.453   0.594   0.525   0.638   0.752
hist(out.corr, main = "", xlab="correlation", ylab="")
```

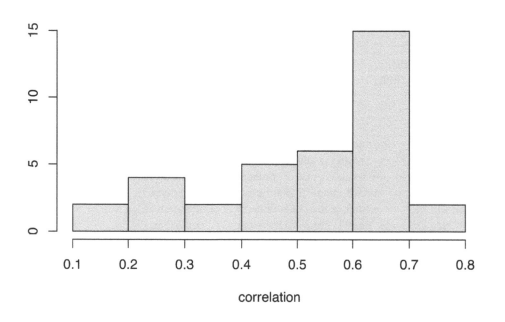

FIGURE 9.7: Histogram of correlations between residuals of TNC and Taxi usage after fitting Model Z1 across taxi zones.

Histograms of the estimated coefficients from the *g* zones (see Figure 9.8) can be used to investigate whether the zones exhibit relatively homogeneous behavior, or whether there are outlying zones, say. Histograms of DIC, WAIC, and MAE are shown in Figure 9.9. These can be used to compare Model Z1 to Models Z2–Z3 (see below).

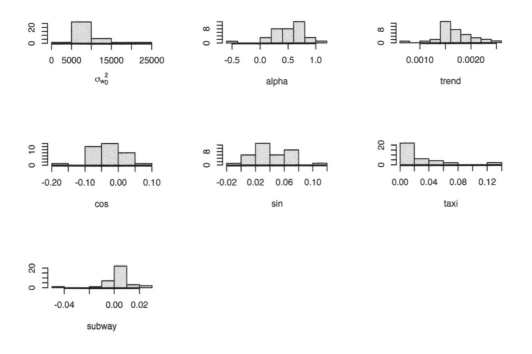

FIGURE 9.8: Histograms of the estimated coefficients from Model Z1 across taxi zones.

Model Z2

Model Z2 is similar to Model Z1, except that the random intercept $\beta_{t,0}$ is modeled as an AR(1) process instead of a random walk, i.e.,

$$\beta_{t,0} = \phi_0 \beta_{t-1,0} + w_{t,0}, \qquad (9.8)$$

where $|\phi_0| < 1$ and $w_{t,0}$ is $N(0, \sigma_{w0}^2)$. The code is similar to the code for Model Z1, except that we replace `f(id.t, model = "rw1", constr = FALSE)` by `f(id.t, model = "ar1")` in the formula. Details on the code and results are available on our GitHub link.

Model Z3

Model Z3 is obtained by dropping the deterministic trend term from Model Z1, so that

$$\log \lambda_t = \alpha + \beta_{t,0} + \sum_{j=1}^{4} \beta_j Z_{t,j}. \qquad (9.9)$$

The code is similar to the code for Model Z1, except that we replace the formula by

```
formula.zone.Z3 <- tnc ~ 1 +  costerm + sinterm + taxi + subway +
    f(id.t, model = "rw1", constr = FALSE)
```

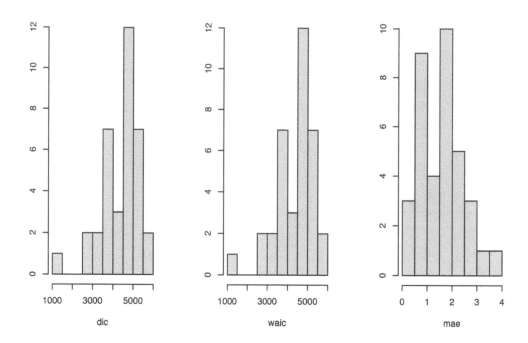

FIGURE 9.9: Histograms of model comparison criteria from Model Z1 across taxi zones.

The results from Model Z1–Z3 show that taxi zones ID127 and ID153 exhibit behavior that is very different than the other $g_1 = 34$ zones. We therefore exclude these zones when we fit the hierarchical model below.

Model Z4. Hierarchical model aggregating across zones

We fit a hierarchical version of Model Z1 with different levels α_i to TNC usage data from $g_1 = 34$ taxi zones, excluding zones ID127 and ID153.

$$y_{it}|\lambda_{it} \sim \text{Poisson}(\lambda_{it}),$$

$$\log \lambda_{it} = \alpha_i + \gamma t + \beta_{it,0} + \sum_{j=1}^{4} \beta_j Z_{it,j}$$

$$\beta_{it,0} = \beta_{i,t-1,0} + w_{it,0}, \tag{9.10}$$

where $w_{it,0} \sim N(0, \sigma_{w0}^2)$.

```
ride.hzone <- ride.zone %>%
  filter(zoneid != "ID127" & zoneid != "ID153")
g1 <- n_distinct(ride.hzone$zoneid)
trend <- 1:n
htrend <- rep(trend, g1)
hcosterm <- rep(costerm, g1)
hsinterm <- rep(sinterm, g1)
```

Model Z4a

The levels α_i are assumed to be fixed coefficients for the $g_1 = 34$ zones. Indexes, replicates, and the data frame for Model Z4a are set up below. The formula includes the levels as fixed effects using `fact.alpha <- as.factor(id.alpha)`.

```
id.b0 <- rep(1:n, g1)
re.b0 <- id.alpha <- rep(1:g1, each = n)
fact.alpha <- as.factor(id.alpha)
ride.hzone.data.fe <-
   cbind.data.frame(ride.hzone, htrend, hcosterm, hsinterm, id.b0, re.b0,
                  fact.alpha)
formula.hzone.fe <-
   tnc ~ -1 + fact.alpha + htrend + hcosterm + hsinterm + taxi + subway +
   f(id.b0, model = "rw1", replicate = re.b0)

model.hzone.fe <- inla(
   formula.hzone.fe,
   data = ride.hzone.data.fe,
   family = "poisson",
   control.predictor = list(compute = TRUE, link = 1),
   control.compute = list(dic = TRUE, waic = TRUE, cpo = TRUE)
)
#summary(model.hzone.fe)
```

We show posterior summaries on a few of the estimated α_i and other fixed effects and random hyperparameters.

```
summary.model.hzone.fe <- summary(model.hzone.fe)
format.inla.out(summary.model.hzone.fe$fixed[30:39,c(1,2)])
##     name          mean    sd
## 1   fact.alpha30  0.906  0.035
## 2   fact.alpha31  1.761  0.035
## 3   fact.alpha32  1.573  0.034
## 4   fact.alpha33  0.844  0.035
## 5   fact.alpha34  1.486  0.034
## 6   htrend        0.002  0.000
## 7   hcosterm     -0.021  0.002
## 8   hsinterm      0.062  0.002
## 9   taxi          0.009  0.000
## 10  subway        0.001  0.000
format.inla.out(summary.model.hzone.fe$hyperpar[,c(1,2)])
##     name                  mean sd
## 1 Precision for id.b0 6767 420.5
```

To see the actual values of the posterior means and standard deviations of the coefficients of Taxi and Subway usage, we can use the code below:

```
tail(model.hzone.fe$summary.fixed[c(1,2)],2)
##              mean          sd
## taxi    0.008904  8.055e-05
## subway  0.001205  6.041e-05
```

Model Z4b

Since there are $g_1 = 34$ zones, we may instead wish to treat the α_i as random effects, i.e., $\alpha_i \sim N(0, \sigma_\alpha^2)$, where σ_α^2 is unknown. The indexes, replicates, formula, and model fit are shown below.

```
id.b0 <- rep(1:n, g1)
re.b0 <- id.alpha <- rep(1:g1, each = n)
prec.prior <-
  list(prec = list(param = c(0.001, 0.001), initial = 20))
ride.hzone.data.re <-
  cbind.data.frame(ride.hzone, htrend, hcosterm, hsinterm, id.b0, re.b0,
                   id.alpha)
formula.hzone.re <-
  tnc ~ -1 + htrend + hcosterm + hsinterm + taxi + subway +
  f(id.alpha, model = "iid", hyper = prec.prior) +
  f(id.b0, model = "rw1", replicate = re.b0)

model.hzone.re <- inla(
  formula.hzone.re,
  family = "poisson",
  data = ride.hzone.data.re,
  control.predictor = list(compute = TRUE, link = 1),
  control.compute = list(
    dic = TRUE,
    cpo = TRUE,
    waic = TRUE,
    config = TRUE,
    cpo = TRUE
  )
)
summary(model.hzone.re)

summary.model.hzone.re <- summary(model.hzone.re)
format.inla.out(summary.model.hzone.re$fixed)
## name      mean   sd     0.025q  0.5q    0.975q  mode    kld
## 1 htrend   0.002  0.000  0.002   0.002   0.002   0.002   0
## 2 hcosterm -0.021 0.002  -0.025  -0.021  -0.017  -0.021  0
## 3 hsinterm 0.062  0.002  0.059   0.062   0.065   0.062   0
## 4 taxi     0.009  0.000  0.009   0.009   0.009   0.009   0
## 5 subway   0.001  0.000  0.001   0.001   0.001   0.001   0
format.inla.out(summary.model.hzone.re$hyperpar[,c(1,2)])
## name                      mean       sd
## 1 Precision for id.alpha   0.575      0.142
## 2 Precision for id.b0      6752.865   425.638
```

Boxplots of α_i's generated from `inla.posterior.sample` for Model Z4b are shown in Figure 9.10. The distribution of the level varies across zones.

In-sample comparison criteria for Models Z1 to Z4b are shown in Table 9.1.

Remark. We can run similar hierarchical models corresponding to Model Z2 and Model Z3 (results not shown here) and compare the different models for TNC usage in the $g_1 = 34$ taxi zones.

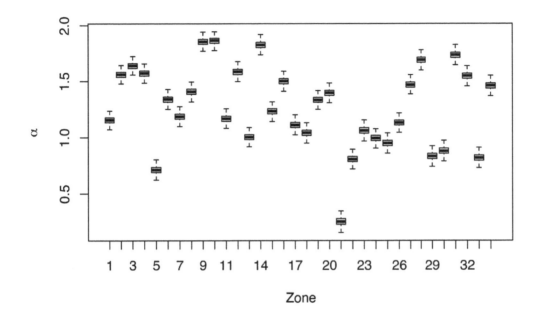

FIGURE 9.10: Boxplots of α_i's for 34 zones from Model Z4b.

TABLE 9.1: In-sample comparisons of Models Z1-Z4b.

Model	DIC	WAIC	Average in-sample MAE
Model Z1	4418	4399	1.604
Model Z2	4422	4403	1.616
Model Z3	4427	4400	1.561
Model Z4a	159074	158597	2.168
Model Z4b	159074	158596	2.168

10

Modeling Stochastic Volatility

10.1 Introduction

An important task in analyzing financial data involves fitting models to study the volatility of asset returns. The class of autoregressive conditionally heteroscedastic (ARCH) models was proposed by Engle (1982), and was extended by Bollerslev (1986) to generalized ARCH (GARCH) models. The simple ARCH(1) model for time series of returns r_t has the form

$$r_t = \sigma_t \varepsilon_t,$$
$$\sigma_t^2 = \alpha_0 + \alpha_1 r_{t-1}^2, \tag{10.1}$$

where σ_t^2 is the conditional variance of r_t given the past history $\{r_{t-1}, r_{t-2}, \ldots\}$, $\alpha_0 > 0$, $0 \leq \alpha_1 < 1$, and ε_t is an i.i.d. process with mean 0 and variance 1. The GARCH(1,1) model generalizes the ARCH(1) model and is given by

$$r_t = \sigma_t \varepsilon_t,$$
$$\sigma_t^2 = \alpha_0 + \alpha_1 r_{t-1}^2 + \beta_1 \sigma_{t-1}^2, \tag{10.2}$$

where $\alpha_0 > 0$, and $\alpha_1 + \beta_1 < 1$. These models are sometimes called *singly stochastic models* in the literature because they only have one error ε_t in the multiplicative model equation for r_t, but no error in the model for the latent conditional variances σ_t^2. The GARCH(m, r) model generalizes these models to include m ARCH and r GARCH coefficients. There is an enormous literature on such models, especially in the econometrics domain.

The stochastic volatility (SV) models that were introduced by Taylor (1982) are alternatives to GARCH models and are widely used for modeling volatility in financial time series. SV models introduce an additional error in the (state) model for the latent conditional variance σ_t^2, and are referred to in the literature as *doubly stochastic models*. This chapter describes a state space formulation for SV modeling of a time series.

We source the custom functions.

```
source("functions_custom.R")
```

This chapter uses the following packages.

```
library(quantmod)
library(INLA)
library(tidyverse)
library(astsa)
library(gridExtra)
```

We analyze the following data.

1. IBM returns from 2018-01-01 to 2020-12-31.

2. NYSE returns from the R package astsa.

10.2 Univariate SV models

Jacquier et al. (1994) described dynamic Bayesian modeling of an SV model. We follow the notation in Martino et al. (2011), who described fitting SV models using R-INLA. The SV model for asset returns r_t is defined by

$$r_t = \exp(h_t/2)\varepsilon_t,$$
$$h_t - \mu = \phi(h_{t-1} - \mu) + \sigma_w w_t, \tag{10.3}$$

where h_t is the unobserved log volatility, ε_t and w_t are i.i.d. $N(0,1)$ random errors which are independent of each other, $|\phi| < 1$, and $\mu \in \mathbb{R}$. The model assumes that the latent log volatility h_t follows a stationary AR(1) process. Following the discussions in Martino et al. (2011), we assume a normal prior with mean 0 and a large variance for μ, an inverse Gamma prior for σ_w^2, while ϕ has the usual logit prior. The following example illustrates the use of the family="stochvol" option for data simulated from an SV model with $\varepsilon_t \sim N(0,1)$.

10.2.1 Example: Simulated SV data with standard normal errors

We simulate a time series following (10.3); see inla.doc("stochvol"). Here, $\varepsilon_t \sim N(0,1)$, as indicated by ret <- exp(h/2) * rnorm(n). The parameter $\phi = 0.9$ generates a series with high persistence.

```
set.seed(1234)
n <- 1000
phi <- 0.9
mu <- -2.0
h <- rep(mu, n)
sigma2.w <- 0.25
prec.x <- prec.state.ar1(sigma2.w, phi)
for (i in 2:n) {
  h[i] <- mu + phi * (h[i - 1] - mu) + rnorm(1, 0, sqrt(sigma2.w))
}
ret <- exp(h/2) * rnorm(n)
```

We use family = "stochvol" to model the SV model. The results are summarized below.

```
time <- 1:n
data.sim.sv <- cbind.data.frame(ret = ret, time = time)
formula.sim.sv <- ret ~ 0 + f(time, model = "ar1",
                        hyper = list(prec = list(param=c(1,0.0001)),
                              mean = list(fixed = FALSE)))
model.sim.g <-
  inla(formula.sim.sv, family = "stochvol", data = data.sim.sv,
```

```
        control.compute = list(
    dic = TRUE,
    waic = TRUE,
    cpo = TRUE,
    config = TRUE),
  )
```
#summary(model.sim.g)
```
format.inla.out(model.sim.g$summary.hyperpar[c(1,2)])
##    name                 mean   sd
## 1 Precision for time   0.949  0.151
## 2 Rho for time                0.888  0.026
## 3 Mean for time               -2.133 0.139
```

Note that the mean μ of the log volatility h_t in the latent AR(1) state equation (10.3) is recovered from the estimate of "Mean for time" in the output, with posterior mean -2.133 and posterior standard deviation 0.139. If there are exogenous predictors, we can replace `model = "ar1"` by `model = "ar1c"`.

10.2.2 Example: Simulated SV data with Student-t_ν errors

The model generating mechanism is the same as in the previous example, except for assuming that $\varepsilon_t \sim t_\nu$, i.e., a Student-t distribution with ν degrees of freedom (d.f.), ν being treated as an unknown parameter. The t-distribution allows for heavier tails than the normal and is widely used in modeling financial data (Tsay, 2005). The use of `stochvol_t` enables us to adequately recover the true values. The *dof-prior* has a tabulated density and is the PC prior for the degrees of freedom parameter in the Student-t distribution (see Section 3.9.2). The parameters of this prior are (u, α) with the interpretation that $P(\nu < u) = \alpha$. The code for data generation is the same as the code for the Gaussian case, except for replacing

```
ret <- exp(h/2) * rnorm(n)
```

by

```
ret <- exp(h / 2) *rt(n, df=5)
```

The code for the formula and model fit are also similar to the Gaussian case, except for replacing `family = "stochvol"` by `family = "stochvol_t"` in the `inla()` call.

```
format.inla.out(model.sim.t$summary.hyperpar[c(1,2,4)])
##    name                                             mean   sd     0.5q
## 1 degrees of freedom for stochvol student-t        12.477 8.800  9.841
## 2 Precision for time                               0.837  0.161  0.826
## 3 Rho for time                                     0.875  0.037  0.881
## 4 Mean for time                                    -1.796 0.162 -1.792
```

10.2.3 Example: IBM stock returns

We download daily prices P_t for IBM using the R package `quantmod`, and compute the returns as $r_t = \nabla \log P_t = (1 - B) \log P_t$, where B is the backshift operator. The IBM price data from 2018-01-01 to 2020-12-31 is shown in Figure 10.1, together with the returns (we dropped the first NA).

```
library(gridExtra)
getSymbols(
  "IBM",
  from = '2018-01-01',
  to = "2020-12-31",
  warnings = FALSE,
  auto.assign = TRUE
)
## [1] "IBM"
ret <- ts(diff(log(IBM$IBM.Adjusted)))[-1]
adj.plot <-
  tsline(
    as.numeric(IBM$IBM.Adjusted),
    xlab = "t",
    ylab = "Adjusted price",
    line.color = "black",
    line.size = 0.6
  )
ret.plot <- tsline(
  ret,
  xlab = "t",
  ylab = "Returns",
  line.color = "red",
  line.size = 0.6
)
grid.arrange(adj.plot, ret.plot, nrow = 1)
```

We fit the univariate SV model discussed in (10.3) to the time series of length $n = 754$ of daily returns for IBM, assuming three different possible distributions for ε_t. The option `family = "stochvol"` fits an SV model (10.3) with i.i.d. N(0,1) errors ε_t.

```
set.seed(1234)
id.1 <- 1:length(ret)
data.ret <- cbind.data.frame(ret, id.1)
formula.ret <- ret ~ 0 + f(id.1,model="ar1",
                    hyper = list(prec = list(param=c(1,0.0001)),
                                 mean = list(fixed = FALSE)))
model.g <- inla(
  formula.ret,
  family = "stochvol",
  data = data.ret,
  control.compute = list(
    dic = TRUE,
    waic = TRUE,
    cpo = TRUE,
    config = TRUE
  ),
  control.predictor = list(compute = TRUE, link = 1)
)
#summary(model.g)
format.inla.out(model.g$summary.hyperpar[c(1,2,4)])
```

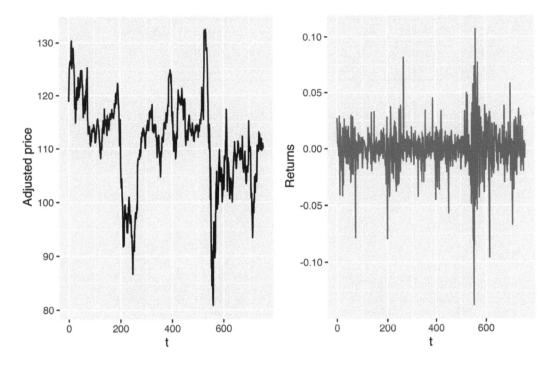

FIGURE 10.1: Adjusted prices (left) and log returns (right) for IBM.

```
##    name                   mean    sd      0.5q
## 1 Precision for id.1   0.038  0.020   0.035
## 2 Rho for id.1          0.998  0.001   0.998
## 3 Mean for id.1        -0.439  0.675  -0.463
```

The estimated posterior mean of ϕ is 0.998, which indicates very high persistence. Note that `summary.fitted.values` gives the fitted values of σ_t^2. We can obtain fitted values of $\exp(h_t)$ using `summary.linear.predictor`. We overlay a plot of the squared returns r_t^2 and the posterior mean of $\exp(h_t)$; see the top panel of Figure 10.2.

```
fit.model.g <- exp(model.g$summary.linear.predictor$mean)
```

If we suspect that the assumption of standard normal errors ε_t may not be suitable, we can also fit a model with i.i.d. Student-t errors by replacing `family = "stochvol"` in the `model.g` call above by `family = "stochvol_t"` in `model.t`. The results are shown below.

```
#summary(model.t)
format.inla.out(model.t$summary.hyperpar[c(1,2)])
##    name                                           mean    sd
## 1 degrees of freedom for stochvol student-t   5.842  1.225
## 2 Precision for id.1                           0.055  0.034
## 3 Rho for id.1                                 0.999  0.001
## 4 Mean for id.1                               -0.541  0.879
```

```
fit.model.t <-
  exp(model.t$summary.linear.predictor$mean)
```

TABLE 10.1: In-sample model comparisons.

Model	DIC	WAIC	Marginal Likelihood
model.g	−4262.23	−4238.848	2070.688
model.t	−4260.187	−4259.245	2084.249

The high value of the estimated ϕ indicates a high degree of persistence. The estimated d.f. for the t-distribution is about 6. Again, the fitted means of $\exp(h_t)$ are obtained as shown earlier for `model.g`. A plot of these values and r_t^2 is shown in the lower plot in Figure 10.2.

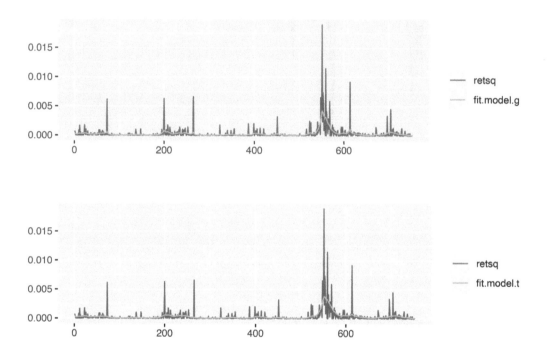

FIGURE 10.2: Comparing observed squared returns (red) and fitted squared volatilities (blue) from the two models for the IBM data.

We note that due to Monte Carlo sampling which `INLA` uses internally, there are likely to be small differences in the results when running `family = stochvol_t` on different computers. But overall, the results should be more or less the same (in terms of reproducibility). Martino et al. (2011) also describe how to use `family = "stochvol_nig"` for SV modeling where ε_t follows a normal inverse Gaussian distribution. See Barndorff-Nielsen (1997) and Andersson (2001) who described the NIG distribution as a variance-mean mixture of a normal distribution with the inverse Gaussian distribution as the mixing distribution. We refer the reader to Martino et al. (2011) for more details. Table 10.1 shows the DIC, WAIC, and the marginal likelihoods for the two models. They are very close, with the values for `model.t` being slightly better.

10.2.4   Example: NYSE returns

To illustrate the approach on a longer time series, we run these codes on the *nyse* data of length $n = 2000$ from the R package **astsa**. We are able to closely recover estimates of σ_w and ϕ from **family = "stochvol_t"**, which were similar to the results in Shumway and Stoffer (2017); however, unlike their setup, we do not include an α in the observation level equation, but only include a state level mean μ.

```r
id.1 <- 1:length(nyse)
data.nyse <- cbind.data.frame(nyse, id.1)
formula.nyse <- nyse ~ 0 + f(id.1,model="ar1",
                       hyper = list(prec = list(param=c(1,0.001)),
                                   mean = list(fixed = FALSE)))
model.nyse <- inla(
  formula.nyse,
  family = "stochvol_t",
  data = data.nyse,
  control.compute = list(
    dic = TRUE,
    waic = TRUE,
    cpo = TRUE,
    config = TRUE
  ),
  control.predictor = list(compute = TRUE, link = 1)
)
#summary(model.nyse)
format.inla.out(model.nyse$summary.hyperpar[,c(1:2)])
##   name                                          mean    sd
## 1 degrees of freedom for stochvol student-t    5.144  0.664
## 2 Precision for id.1                           2.269  0.984
## 3 Rho for id.1                                 0.994  0.005
## 4 Mean for id.1                               -9.048  0.335
```

11

Spatio-temporal Modeling

11.1 Introduction

When data is available over space and time, we can use models that can capture both spatial and temporal (random) effects, as well as interaction effects between space and time. A spatio-temporal model is said to be separable when the space-time covariance structure can be decomposed into a spatial and a temporal term, i.e., the spatio-temporal covariance can be expressed as a Kronecker product of a spatial and a temporal covariance (Fuentes et al., 2008). Non-separable covariance structures do not allow for separate modeling of the spatial and temporal covariances (Wikle et al., 1998). Spatial and space-time models can be fitted using INLA, as discussed in several papers including Lindgren and Rue (2015), Bakka et al. (2018), Krainski et al. (2018), and Van Niekerk et al. (2019). In this chapter, we describe fitting dynamic spatio-temporal models to ridesourcing data in NYC using R-INLA, as discussed in Knorr-Held (2000) and Blangiardo et al. (2013).

As in previous chapters, the custom functions must be sourced.

```
source("functions_custom.R")
```

The list of packages used in this chapter is shown below. Note that we have sourced `tidyverse` as the last package here, in order to avoid some of its functions to be overridden by similar functions in packages related to spatial analysis.

```
library(lubridate)
library(gridExtra)
library(raster)
library(tmap)
library(sf)
library(spdep)
library(mapview)
library(gtools)
library(INLA)
library(tidyverse)
```

We analyze the following data.

1. Monthly TNC usage in NYC.

11.2 Spatio-temporal process

A spatial process in d dimensions can be written as

$$\{Y(\boldsymbol{s}) : \boldsymbol{s} \in D \subset \mathcal{R}^d\},$$

where $d = 2$, $\boldsymbol{s}$ indexes the location, and $Y(\boldsymbol{s})$ represents the variable of interest at location $\boldsymbol{s}$. Spatial data can either be areal (lattice) data, or point referenced (geostatistical) data (Cressie, 2015). For areal data, the domain D can be a regular or irregular lattice, and is partitioned into a finite number of areal units with well-defined boundaries, such as ZIP codes, census tracts, taxi zones, etc. By contrast, for geostatistical data, D is assumed to be a continuous fixed set, so that $\boldsymbol{s}$ is deterministic and can vary continuously over D, and $Y(\boldsymbol{s})$ is observable at any point within D, such as weather measurements at monitoring stations.

With areal spatial data, the relationship between g regions is characterized in terms of adjacency, and observations from neighboring regions may exhibit higher association than those from distant regions. The neighborhood relationship $i \sim j$ is defined in terms of a $g \times g$ adjacency weight matrix $\boldsymbol{W}$, with positive entries $w_{i,j}$ and $w_{j,i}$ at the intersection of rows and columns of neighboring areas, and zero entries otherwise.

We define a *spatio-temporal process* in $\mathcal{R}^2 \times \mathcal{R}$ by

$$\{Y(\boldsymbol{s}, t) : \boldsymbol{s} \in D \subset \mathcal{R}^2, t \in \mathcal{R}\},$$

where $\boldsymbol{s}$ indexes the location, t indexes time, and $Y(\boldsymbol{s}, t)$ represents a variable of interest. We focus on areal data and assume that the spatio-temporal process is observed at g areal regions over n time points. An areal data set may be represented as a *shapefile*, i.e., in a data format that stores the location, shape, and attributes of spatial features such as points, lines, and polygons (Moraga, 2019).

11.3 Dynamic spatial models for areal data

Knorr-Held (2000) described dynamic spatio-temporal models as follows; also see Blangiardo et al. (2013), and Moraga (2019).

Model 1.

Let $\lambda_{it} = E(Y(\boldsymbol{s}_i, t))$; then $\eta_{it} = \log(\lambda_{it})$ can be modeled as

$$\eta_{it} = \alpha + \nu_i + \epsilon_i + \gamma_t + \phi_t, \tag{11.1}$$

where

    (i) α denotes an overall intercept (across all $\boldsymbol{s}_i$ and t),

    (ii) ν_i is a spatially structured effect specific to areal unit i,

    (iii) $\epsilon_i \sim N(0, \sigma_\epsilon^2)$ is an unstructured error corresponding to areal unit i,

    (iv) γ_t is a temporally structured effect with a random walk evolution, and

(v) $\phi_t \sim N(0, \sigma_\phi^2)$ is an unstructured error corresponding to time t.

The effect ν_i can be modeled by an intrinsic conditional autoregressive (ICAR) specification

$$\nu_i | \nu_j \sim N(m_i, M_i^2), \text{for } j \neq i, \text{ where,}$$

$$m_i = \frac{\sum_{j \in \mathcal{N}(i)} \nu_j}{card(\mathcal{N}(i))},$$

$$M_i^2 = \frac{\sigma_\nu^2}{card(\mathcal{N}(i))}, \tag{11.2}$$

where $\mathcal{N}(i)$ denotes the set of neighbors of s_i, and $card(\mathcal{N}(i))$ is its cardinality. The temporally structured effect γ_t has a random walk evolution defined as

$$\gamma_t | \boldsymbol{\gamma}_{(-t)} \sim \begin{cases} N(\gamma_{t+1}, \tau_\gamma) & t = 1, \\ N(\frac{1}{2}(\gamma_{t-1} + \gamma_{t+1}), \frac{1}{2}\tau_\gamma) & t = 2, \ldots, n-1, \\ N(\gamma_{t-1}, \tau_\gamma) & t = n, \end{cases} \tag{11.3}$$

where, $\boldsymbol{\gamma}_t = (\gamma_1, \ldots, \gamma_n)'$, and $\boldsymbol{\gamma}_{(-t)}$ denotes $\boldsymbol{\gamma}_t$ excluding γ_t.

Model 2.

Model 2 extends Model 1 by allowing a space-time interaction term δ_{it} in (11.1) as

$$\eta_{it} = \alpha + \nu_i + \epsilon_i + \gamma_t + \phi_t + \delta_{it}, \tag{11.4}$$

where all other terms are defined as in Model 1. There are four ways to define δ_{it}, depending on which of the spatial terms ν_i and ϵ_i interact with one of the temporal effects, γ_t or ϕ_t (Blangiardo et al., 2013). Their descriptions involve the ICAR models with the *bym*, or *besag*, or *besagproper* specifications (see Besag (1974) or Besag et al. (1991)).

11.4   Example: Monthly TNC usage in NYC taxi zones

In Chapter 6, we discussed temporal models for weekly TNC usage from Jan 1, 2015–June 30, 2017 across several taxi zones in NYC, while in Chapter 9, we modeled daily TNC usage for $g = 36$ zones in Manhattan using dynamic models with Poisson based observation equations. In those cases, we ignored possible spatial associations between the taxi zones.

Here, we fit separable spatio-temporal models to monthly TNC counts (in 1000's), for $g = 252$ zones in NYC. We constructed this data by dropping taxi zones with no TNC usage for any of the $n = 30$ months. In Model 1, the spatial effect is explained using an ICAR model (see (11.2)), while the temporal trend is additive and is handled by a latent random walk (see (11.3)). We then discuss different ways to incorporate the interaction δ_{it} under Model 2.

Data preprocessing

While the data *ride.3yr* is arranged by zone, the data matrix *ride* is arranged by year and month, using the format $\boldsymbol{D} = (\boldsymbol{D}_1', \ldots, \boldsymbol{D}_n')'$, where $\boldsymbol{D}_t'$ is a $g \times 6$ matrix for each $t = 1, \ldots, n$. The six columns correspond to *year, month, zoneid, zone.name, borough,* and *tnc.month*. Note that we have used the `mixedorder()` function from the `gtools` package to arrange the zoneid numerically.

```
ride.3yr <-
  read.csv("tnc_monthly_data_for_spatiotemporal.csv",
           header = TRUE,
           stringsAsFactors = FALSE)
ride <- ride.3yr[mixedorder(ride.3yr$zoneid),] %>%
  arrange(year, month)
ride$month.year <- paste(ride$year, ride$month, sep="-")
n <- n_distinct(ride$month.year)
g <- n_distinct(ride$zoneid)
```

We use the following code which allows us to read the *zoneshape* file for all available taxi zones in NYC that we obtained with the rideshare data. We merge the two data frames *zoneshape* and *ride* by the common *zoneid* column. Note that information on shape is given via a collection of files that have different extensions, a common name, and are stored in the same directory. To run these models, the following files must be saved in the same folder as the data: taxi_zones.dbf, taxi_zones.prj, taxi_zones.sbn, taxi_zones.sbx, taxi_zones.shp, taxi_zones.shp.xml, and taxi_zones.shx. Here, .shp contains the geometry data, .shx is a positional index of the geometry data that allows us to seek forward and backward in the .shp file, and .dbf stores the attributes for each shape.

```
zoneshape <- st_read("taxi_zones.shp",quiet = TRUE)
zoneshape$zoneid <- paste("ID", zoneshape$LocationID, sep = "")
match.zone <- match(unique(ride$zoneid), zoneshape$zoneid)
zoneshape <- zoneshape[match.zone, ] %>%
  arrange(OBJECTID)
```

The `poly2nb()` function provides an adjacency matrix for the NYC taxi zones with 1's or 0's. It has an argument `queen`, which constructs queen adjacency when `queen=TRUE` (default setup).

```
nb.q <- poly2nb(zoneshape, queen=TRUE,row.names = zoneshape$zoneid)
```

We transform an `nb` object into an adjacency matrix *nyc.adj* using the `nb2INLA()` function.

```
nb2INLA("map.adj", nb.q)
nyc.adj <- inla.read.graph(filename = "map.adj")
```

Remark. A rook adjacency is constructed when `queen=FALSE` (not computed here).

```
nb.r <- poly2nb(ride.inner.tnc, queen=FALSE)
```

We define the following indexes to use with Model 1 and Model 2 with four different types of interactions. In the code below, *id.zone* denotes the structured spatial effect ν_i, while *id.month* and *id.monthu* respectively denote the structured time effect γ_t and the unstructured time effect ϕ_t. The indexes *id.zone.int*, *id.month.int*, and *id.zoneu.monthu* enable us to set up the interaction effects in the different models, as described below. We first construct a factor variable *zoneid* with as many levels as there are unique zones in the data set. Then, *id.zone* and *id.zone.int* are set up as `as.numeric(zoneid)`, and each contains the sequence $1, 2, \ldots, g$, with each value in the sequence (representing a zone) repeated n times. For month effects, we have the sequence $1, \ldots, n$ repeated g times in the indexes *id.month*, *id.monthu*, and *id.month.int*. The last index is *id.zoneu.monthu* which assumes values from 1 to $n \times g$.

```
zoneid <- factor(ride$zoneid, levels=unique(zoneshape$zoneid))
id.zone <- id.zone.int <- as.numeric(zoneid)
id.month <- id.monthu <- id.month.int <- rep(1:n, each = g)
id.zoneu.monthu <- 1:nrow(ride)
```

Before running the models, it is wise to ascertain that the number and order of the zones are the same in the *ride* data and *zoneshape* data.

```
identical(unique(ride$zoneid), zoneshape$zoneid)
## [1] TRUE

data.kh <- cbind.data.frame(ride,
                            id.zone, id.zone.int,
                            id.month,
                            id.monthu,
                            id.month.int,
                            id.zoneu.monthu)
```

11.4.1 Model 1. Knorr-Held additive effects model

We fit the Knorr-Held model (Model 1) to TNC counts, with additive spatial and temporal effects defined in (11.1)–(11.3). We stack the time series for each month, one below the other. Since the formula does not have to explicitly include a function to model the unstructured spatial effect ϵ_i, we do not include this variable (which would be denoted by *id.zoneu*) in the data frame. The formula and model under the *bym* prior are shown below.

```
formula.add <- tnc.month ~ 1 + f(id.zone, model = "bym", graph = nyc.adj) +
  f(id.month, model = "rw1") +
  f(id.monthu, model = "iid")

tnc.add <- inla(
  formula.add,
  data = data.kh,
  control.predictor = list(compute = TRUE, link = 1),
  control.compute = list(dic = TRUE, waic = TRUE, cpo = TRUE),
  family = "poisson"
)
format.inla.out(summary.tnc.add$fixed[,c(1:2), drop=FALSE])
format.inla.out(summary.tnc.add$hyperpar[,c(1:2)])
```

```
##    name         mean  sd
## 1 Intercept    2.244 0.079
##    name                                    mean       sd
## 1 Precision for id.zone iid component      0.659      0.068
## 2 Precision for id.zone spatial component  2.606      1.122
## 3 Precision for id.month                 141.401     37.443
## 4 Precision for id.monthu              24394.090  21634.086
```

The fixed effect corresponds to α in (11.1). The random effects correspond, in order, to the unstructured spatial effect ϵ_i, structured spatial effect ν_i, unstructured time effect ϕ_t, and structured time effect γ_t.

We can explore the output in greater detail beyond the summary. The dimension of `summary.tnc.khadd$random$id.zone` is $2g \times 8$. The first g rows correspond to $\nu_i + \epsilon_i$, the posterior summary of the spatial random effect, i.e., the *sum* of structured spatial and i.i.d. spatial effects. The last g rows contains the posterior summary for the structured spatial effect ν_i.

Remark. Instead of the *bym* prior, we can use either the *besag* or *besagproper* priors for the structured spatial effect (results not shown here).

11.4.2   Knorr-Held models with space-time interactions

We describe four different models to describe the interaction δ_{it} in Model 2 defined in (11.4).

Model KH1. Knorr-Held model with interaction between ϵ_i and ϕ_t

In Model KH1, the term δ_{it} represents an interaction between the unstructured spatial effect ϵ_i and the unstructured temporal effect ϕ_t. We assume that $\delta_{it} \sim N(0, \sigma_\delta^2)$. In the code below, the index for the interaction term is *id.zoneu.monthu*, which is a series of integers from 1 to $n \times g$, which is 7560. This is represented by the term `f(id.zoneu.monthu, model = "iid")` in the formula. Since the interaction is between two unstructured effects, the `iid` model is used to estimate σ_δ^2.

```
formula.kh1 <- tnc.month ~ 1 + f(id.zone, model = "bym", graph = nyc.adj) +
  f(id.month, model = "rw1") +
  f(id.monthu, model = "iid") +
  f(id.zoneu.monthu, model = "iid")

tnc.kh1 <- inla(
  formula.kh1,
  data = data.kh,
  control.predictor = list(compute = TRUE, link = 1),
  control.compute = list(dic = TRUE, waic = TRUE, cpo = TRUE),
  family = "poisson"
)
format.inla.out(summary.tnc.kh1$fixed[,c(1:2), drop=FALSE])
format.inla.out(summary.tnc.kh1$hyperpar[,c(1:2)])
```

```
##    name       mean  sd
## 1 Intercept 2.217 0.078
##    name                                      mean         sd
## 1 Precision for id.zone iid component         0.663      0.066
## 2 Precision for id.zone spatial component     2.153      0.638
## 3 Precision for id.month                    153.004     27.734
## 4 Precision for id.monthu                 18655.012  11340.821
## 5 Precision for id.zoneu.monthu              76.871      3.747
```

The fixed effect corresponds to α in Model 2. The first four random effects are what we had in the additive model. The last effect corresponds to the interaction δ_{it}.

Model KH2. Knorr-Held model with interaction between ν_i and ϕ_t

This model describes an interaction between the structured spatial effect ν_i denoted by the *besag* specification (indexed by *id.zone*), and the unstructured temporal effect ϕ_t (indexed by *id.monthu*). We specify a conditional autoregressive (CAR) framework for the zones for each week independently from all the other; thus, for the index *id.month.int* which represents the unstructured temporal effect, the model is "iid" for each zone (see `group = id.zone.int` below).

Note that if the interaction δ_{it} involves either a *structured* spatial or a *structured* temporal effect (as in Models KH2–KH4), we must include `group` and `control.group` in the formula, as shown below. The `inla()` call to obtain `tnc.type2` is similar to the code for Model KH1, except for replacing `formula.type1` by `formula.type2`. The output is shown below.

```
formula.kh2 <- tnc.month ~ 1 +
  f(id.zone, model = "bym", graph = nyc.adj) +
  f(id.month, model = "rw1") +
  f(id.monthu, model = "iid") +
  f(
    id.month.int,
    model = "iid",
    group = id.zone.int,
    control.group = list(model = "besag", graph = nyc.adj)
  )
```

```
##   name        mean  sd
## 1 Intercept 0.762 0.079
##   name                                      mean     sd
## 1 Precision for id.zone iid component       1.432    0.219
## 2 Precision for id.zone spatial component   1.284    0.380
## 3 Precision for id.month                  240.020  198.785
## 4 Precision for id.monthu                   4.752    2.316
## 5 Precision for id.month.int              101.716    6.542
```

Model KH3. Knorr-Held model with interaction between ϵ_i and γ_t

This model describes an interaction between the unstructured spatial effect ϵ_i and the structured temporal effect γ_t (indexed by *id.month*) by assuming an "rw1" specification for each zone independently from all the other zones. For the index *id.zone.int* which represents the unstructured spatial effect, the model is "iid" indicating independent zones, with each weekly time series within each zone modeled as "rw1". The output is shown below.

```
formula.kh3 <- tnc.month ~ 1 +
  f(id.zone, model = "bym", graph = nyc.adj) +
  f(id.month, model = "rw1") +
  f(id.monthu, model = "iid") +
  f(
    id.zone.int,
    model = "iid",
    group = id.month.int,
    control.group = list(model = "rw1")
  )
```

```
##   name        mean  sd
## 1 Intercept 0.612 0.032
##   name                                      mean     sd
## 1 Precision for id.zone iid component       2.286    0.314
## 2 Precision for id.zone spatial component  14.389    7.090
## 3 Precision for id.month                  207.485   94.423
```

```
## 4 Precision for id.monthu                      2.487  0.667
## 5 Precision for id.zone.int                    56.445  2.830
```

Model KH4. Knorr-Held model with interaction between ν_i and γ_t

Here, the model describes an interaction between the structured spatial effect ν_i denoted by the *besag* specification (indexed by *id.zone*), and the structured temporal effect γ_t (indexed by *id.month*). This assumes a random walk across time for each zone that depends on the neighboring zones. The index *id.zone.int* represents the dependent zones, modeled as *besag*, with each time series within a zone modeled as "rw1".

```
formula.kh4 <- tnc.month ~ 1 +
  f(id.zone, model = "bym", graph = nyc.adj) +
  f(id.month, model = "rw1") +
  f(id.monthu, model = "iid") +
  f(
    id.zone.int,
    model = "besag",
    graph = nyc.adj,
    group = id.month.int,
    control.group = list(model = "rw1")
  )
```

```
##    name       mean  sd
## 1 Intercept 2.082 0.023
##    name                                          mean       sd
## 1 Precision for id.zone iid component            16.41  0.000e+00
## 2 Precision for id.zone spatial component      1035.79  5.120e+02
## 3 Precision for id.month                        102.39  4.962e+01
## 4 Precision for id.monthu                     76867.92  1.017e+05
## 5 Precision for id.zone.int                      68.86  4.257e+00
```

We summarize results from all five fitted models. The intercept α for the K-H additive model and Model KH1 are estimated with a similar posterior mean of about 2.22, while the value is closer to 0.76 in Models KH2–KH4. The scales of the estimated posterior precisions for each of the unstructured and structured spatial and temporal components are quite similar between the models (see the formatted output under each fit). Table 11.1 presents the DIC, WAIC, and in-sample average MAE (across zones) values from the fitted models. The DIC, WAIC, and MAE are smallest for Model KH3. For comparison, we also fitted a model without any spatial effects in Model 1, and its criteria are shown on the last row of Table 11.1. The DIC, WAIC, and average MAE values for this model are very high. Clearly, including spatial effects with or without interactions, is useful.

TABLE 11.1: In-sample model comparisons.

Model	DIC	WAIC	Average MAE
KH additive	38198.723	38254	3.447
KH1	37335.445	37163	2.258
KH2	33585.815	33056	1.392
KH3	33181.453	32417	0.995
KH4		51245	1.041
Temporal model	289664.627	290914	25.044

Figure 11.1 shows the time-averaged residuals across zones from all the models. The code to construct the plot is shown for Model KH3; code for other models is similar. The residuals from Model KH3 are the smallest in almost all zones.

```
res.type3 <- df.type3 %>%
  mutate(obs = data.kh$tnc.month) %>%
  mutate(res = (obs - fit)) %>%
  select(zoneid, res) %>%
  group_by(zoneid) %>%
  summarise(res = mean(res))
res.type3 <- left_join(zoneshape, res.type3, by = "zoneid")
plot.res.type3 <-
  tm_shape(res.type3) + tm_polygons(style = "quantile", col = "res") +
  tm_layout(title = 'Type3')
```

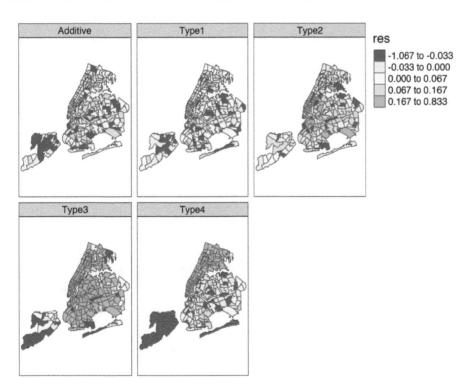

FIGURE 11.1: Time averaged residuals across zones from different models.

Next, we study the behavior of the posterior distributions of ν_i and δ_{it} under Model KH3. Using `inla.posterior.sample()`, we generate 500 samples from the posterior distributions of ν_i and δ_{it} from Model KH3. The top panel in Figure 11.2 shows parallel boxplots of the ν_i values across $g = 252$ zones, while the bottom panel shows boxplots of time averages of the δ_{it} samples. While these boxplots are centered about zero for several zones, we also see non-zero effects in a few taxi zones. The code to obtain the posterior samples is shown below.

```
post.sampletype3 <-
  inla.posterior.sample(n = 500, tnc.type3, seed = 1234)
temp.nu <- temp.delta <- list()
for (k in 1:length(post.sampletype3)) {
  x <- post.sampletype3[[k]]
```

```
temp.nu[[k]] <-
  cbind.data.frame(nu = tail(x$latent[grep("id.zone:",
                                            rownames(x$latent))], g),
                   zone = unique(data.kh$zoneid))
temp.delta[[k]] <-
  cbind.data.frame(delta = x$latent[grep("id.zone.int:",
                                          rownames(x$latent))],
                   zone = data.kh$zoneid) %>%
  group_by(zone) %>%
  summarize(across(delta,  ~ mean(.x)))
}
```

We can use the code below to construct parallel boxplots for the ν_i across $g = 252$ zones. The code for the time-averaged δ_{it} is similar and is not shown here.

```
nu.all <- bind_rows(temp.nu)
nu.all$zone <- factor(nu.all$zone, levels = unique(data.kh$zoneid))
p1 <- ggplot(nu.all, aes(x = zone, y = nu)) +
  geom_boxplot(outlier.shape = NA) +
  ylab(expression(nu)) +
  xlab("zone") +
  theme(axis.text.x = element_blank())
```

Figure 11.3 compares the time averaged (over $n = 30$ months) fitted values from Model KH3 (right plot) with actual monthly TNC usage (left plot) for $g = 252$ zones. The model gives a very close fit.

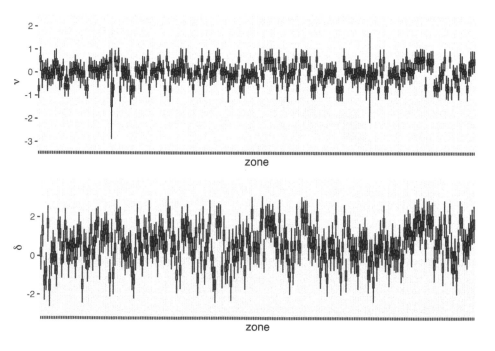

FIGURE 11.2: Boxplots of ν_i (top panel) and boxplots of δ_{it} averaged over time (bottom panel) for 252 zones.

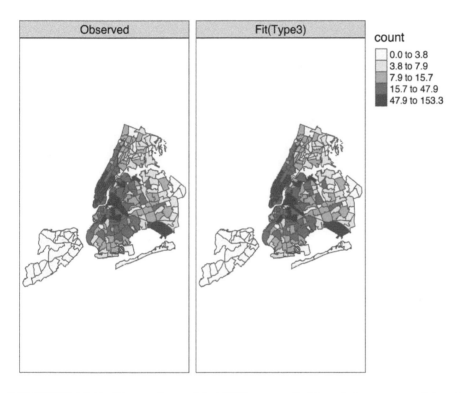

FIGURE 11.3: Observed monthly TNC counts (left) and time averaged fitted counts from Model KH3 (right) for g=252 zones.

12

Multivariate Gaussian Dynamic Modeling

12.1 Introduction

This chapter describes dynamic Bayesian modeling for Gaussian multivariate time series. For $t = 1, \ldots, n$, let $\boldsymbol{y}_t = (Y_{1,t}, \ldots, Y_{q,t})'$ be a q-dimensional time series that is linearly related to a latent p-dimensional process $\boldsymbol{x}_t = (X_{1,t}, \ldots, X_{p,t})'$, plus a q-dimensional vector of Gaussian errors $\boldsymbol{v}_t = (v_{1,t}, \ldots, v_{q,t})'$ with mean $\boldsymbol{0}$ and covariance $\boldsymbol{V}$. For $t = 1, \ldots, n$, the observation equation is

$$\boldsymbol{y}_t = \boldsymbol{F}_t \boldsymbol{x}_t + \boldsymbol{v}_t, \tag{12.1}$$

where the observation error vector $\boldsymbol{v}_t \sim N_q(\boldsymbol{0}, \boldsymbol{V})$ and $\boldsymbol{F}_t$ is a known, measurement matrix (similar to the design/regression matrix in a static linear model). The state vector $\boldsymbol{x}_t$ is assumed to evolve as a first-order vector AR process, i.e., a VAR(1) process. The state equation is

$$\boldsymbol{x}_t = \boldsymbol{\Phi} \boldsymbol{x}_{t-1} + \boldsymbol{w}_t, \tag{12.2}$$

where the state error vector $\boldsymbol{w}_t \sim N_p(\boldsymbol{0}, \boldsymbol{W})$, and $\boldsymbol{\Phi} = \{\phi_{ij}\}$ is a $p \times p$ state transition matrix. We described these models with additional exogenous predictors in (1.26) and (1.27). For more details on the Gaussian DLM and model fitting using the R package `astsa`, see Shumway and Stoffer (2017). Multivariate models can also be implemented using software such as WinBUGS or STAN.

In this chapter, we describe how to use R-INLA to fit a multivariate (MV) DLM in the following situations. In all cases, the observation errors are assumed to be correlated, so that $\boldsymbol{V}$ is a non-diagonal p.d. matrix. We discuss these cases in the following sections.

(a) The state variables follow independent AR(1) processes; i.e., $\boldsymbol{W}$ and $\boldsymbol{\Phi}$ are diagonal matrices (see Section 12.2).

(b) The state errors are equicorrelated, so that $\boldsymbol{W}$ has an intra-class covariance structure. Further, the AR(1) models for each state component have the same coefficient ϕ, so that the matrix $\boldsymbol{\Phi}$ in the VAR(1) model is diagonal with a common ϕ value along the diagonal (see Section 12.3).

(c) The state errors are uncorrelated so that $\boldsymbol{W}$ is diagonal, while the matrix Φ is a full $p \times p$ matrix accommodating dependence in the state vector (see Section 12.4).

Custom functions must be sourced by the user.

```
source("functions_custom.R")
```

We use the following packages.

```
library(INLA)
library(Matrix)
library(mvtnorm)
library(marima)
library(vars)
library(matrixcalc)
library(tidyverse)
```

We analyze the following data.

1. Ridesourcing in NYC

12.2   Model with diagonal W and Φ matrices

We assume that the state error covariance matrix W and the state transition matrix Φ are diagonal matrices. However, the observation error covariance matrix V is a non-diagonal, symmetric p.d. matrix. We first show a simulated example of a bivariate DLM, followed by an illustration using the ridesourcing data for a single taxi zone in NYC.

12.2.1   Description of the setup for V

Suppose that for all t, the observation errors $v_{j,t} \sim N(0, V_{jj})$, $j = 1, \ldots, q$, and that $Corr(v_{j,t}, v_{\ell,t}) = 1$ if $j = \ell$ and $R_{j\ell}$ if $j \neq \ell, j, \ell = 1, \ldots, q$. Let $Cov(v_t) = V = \{V_{j\ell}\}$ and $Corr(v_t) = R = \{R_{j\ell}\}$. Let $D_v = \text{diag}(V_{11}, \ldots, V_{qq})$. Here, V, R, and D_v are $q \times q$ symmetric p.d. matrices, and $R = D_v^{-1/2} V D_v^{-1/2}$. For $q = 2$,

$$V = \begin{pmatrix} V_{11} & V_{12} \\ V_{12} & V_{22} \end{pmatrix} = \begin{pmatrix} \frac{1}{\tau_{11}} & \frac{R_{12}}{\sqrt{\tau_{11}\tau_{22}}} \\ \frac{R_{12}}{\sqrt{\tau_{11}\tau_{22}}} & \frac{1}{\tau_{22}} \end{pmatrix},$$

where the marginal precisions of $v_{1,t}$ and $v_{2,t}$ are $\tau_{11} = 1/V_{11}$ and $\tau_{22} = 1/V_{22}$, while the correlation between $v_{1,t}$ and $v_{2,t}$ is $R_{12} = V_{12}/\sqrt{V_{11}V_{22}}$. In R-INLA, the hyperparameters are τ_{11}, τ_{22}, and R_{12}. The precision matrix V^{-1} is assumed to follow a Wishart distribution, denoted as $V^{-1} \sim \mathcal{W}(r, \Sigma^{-1})$, and we say that V follows an inverse Wishart distribution with d.f. r and scale matrix Σ^{-1}. The "iid2d" option enables us to handle this for augmented structures in multivariate models, following discussion similar to Section 3.3.8 from Gómez-Rubio (2020).

12.2.2   Example: Simulated bivariate AR(1) series

Assume that the observation and state equations are

$$\begin{aligned} y_t &= x_t + v_t, \\ x_t &= \Phi x_{t-1} + w_t. \end{aligned}$$

(12.3)

For the simulation, we first set up observation error and state error covariance matrices.

```
# Observation error covariance V
V.11 <- 1.0
V.22 <- 1.0
R.12 <- 0.7
V.12 <- R.12 * sqrt(V.11) * sqrt(V.22)
V <- matrix(c(V.11, V.12, V.12, V.22), nrow = 2)
# State error covariance W
W.11 <- 0.5
W.22 <- 0.5
W.12 <- 0
W <- matrix(c(W.11, W.12, W.12, W.22), nrow = 2)
```

Since the true **Φ** matrix is diagonal, we can compute the precision of the state variables as

```
phi.1 <- 0.8
phi.2 <- 0.4
prec.x1 <- (1 - phi.1 ^ 2) / W.11
prec.x2 <- (1 - phi.2 ^ 2) / W.22
```

We generate $n = 1000$ states, each from an AR(1) model.

```
set.seed(123457)
n <- 1000
x1 <- arima.sim(n = 1000, list(order = c(1, 0, 0), ar = phi.1),
                sd = sqrt(W.11))
x2 <- arima.sim(n = 1000, list(order = c(1, 0, 0), ar = phi.2),
                sd = sqrt(W.22))
```

Next, we generate $N(\mathbf{0}, \mathbf{V})$ observation errors, and create the bivariate time series $\mathbf{y}_t$.

```
err.v <- rmvnorm(n, mean = c(0, 0), sigma = V)
err.v1 <- err.v[, 1]
err.v2 <- err.v[, 2]
y1 <- x1 + err.v1
y2 <- x2 + err.v2
y <- c(y1, y2)
```

The CCF between $y_{1,t}$ and $y_{2,t}$ shows a non-zero value at lag 0; see Figure 12.1.

```
ccf(y1,y2, main="", ylim = c(-1,1),ylab="ccf",xlab="lag")
```

We show the formula for (12.3), which involves setting up indexes id.V, id.b0.x1, and id.b0.x2. Since R-INLA would fit a univariate Gaussian model by default, we must indicate the requirement for a bivariate Gaussian model (i.e., $q = 2$) by setting model="iid2d" in the formula chunk for id.V. That is, the term f(id.V, model = "iid2d", n = N) indicates that **V** is assumed to be a non-diagonal covariance matrix for the bivariate responses, and follows an inverse Wishart prior.

In R-INLA, "iid2d" is represented internally as a single vector of length $N = 2n$, i.e., $(y_{1,1}, \ldots, y_{1,n}, y_{2,1}, \ldots, y_{2,n})$; N is a required argument in the functional form for id.V. Since R-INLA only allows us to specify up to $q = 5$, the maximum allowed value of N is $N = 5n$, and we can correspondingly set up model = "iid5d". For more details, refer to inla.doc("iid").

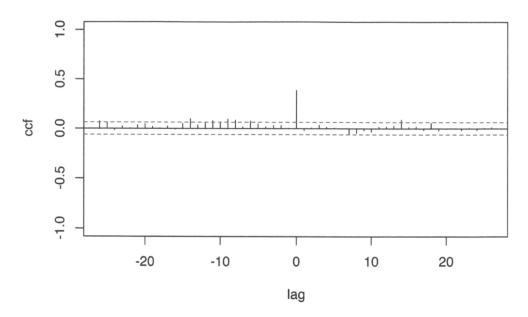

FIGURE 12.1: Sample CCF of the simulated Gaussian bivariate responses with diagonal W and Φ.

The terms `f(id.b0.x1, model = "ar1")` and `f(id.b0.x2, model = "ar1")` indicate how to fit independent AR(1) models to each of the state variables, which are independent because we assumed both Φ and W to be diagonal matrices. We assume a Gaussian prior with high precision through the argument `control.family` (recall that `control.family` is used to define priors on the distribution of the response variable). We can verify from the results that the estimated values are close to the corresponding true values used in the simulation. The estimated level α is close to zero (posterior mean is -0.059 and posterior standard deviation is 0.053), since we did not include a level in the simulated model.

```
m <- n - 1
N <- 2 * n
id.V <- 1:N
id.b0.x1 <- c(1:n, rep(NA, n))
id.b0.x2 <- c(rep(NA, n), 1:n)
biv.dat <- cbind.data.frame(y, id.b0.x1, id.b0.x2, id.V)
```

The formula and model fit are shown below.

```
formula.biv <-  y ~ 1 +
  f(id.V, model = "iid2d", n = N) +
  f(id.b0.x1, model = "ar1") +
  f(id.b0.x2, model = "ar1")

result.biv <- inla(
  formula.biv,
  family = c("gaussian"),
  data = biv.dat,
```

```
  control.inla = list(h = 1e-5, tolerance = 1e-3),
  control.compute = list(dic = TRUE, config = TRUE),
  control.predictor = list(compute = TRUE),
  control.family = list(hyper = list(prec = list(
    initial = 12, fixed = TRUE
  )))
)
result.biv.summary <- summary(result.biv)

format.inla.out(result.biv.summary$hyperpar[c(1,2)])
##    name                                 mean  sd
## 1 Precision for id.V component 1 1.072 0.045
## 2 Precision for id.V component 2 0.686 0.027
## 3 Rho1:2 for id.V                       0.635 0.015
## 4 Precision for id.b0.x1               0.758 0.097
## 5 Rho for id.b0.x1                     0.767 0.016
## 6 Precision for id.b0.x2               4.287 0.980
## 7 Rho for id.b0.x2                     0.587 0.060
```

In the hyperparameter summary, `Precision for id.V (component 1)` estimates the precision $1/V_{11}$, `Precision for id.V (component 2)` estimates the precision $1/V_{22}$, and `Rho1:2 for id.V` estimates the correlation R_{12}. The estimates of the precisions for the state errors are given by `Precision for id.b0.x1` and `Precision for id.b0.x2`, while the estimates for ϕ_1 and ϕ_2 are given by `Rho for id.b0.x1` and `Rho for id.b0.x2`.

12.2.3   Example: Ridesourcing data in NYC for a single taxi zone

We reconsider the ridesourcing data in NYC that we discussed in Chapter 6, where we treated scaled weekly TNC usage as a response variable and scaled Subway and Taxi usage as predictors, along with other time and location varying predictors. We now illustrate the multivariate DLM fit by treating scaled TNC and Taxi usage as a $q = 2$-dimensional *response vector* and fit the model shown in (12.1) and (12.2) to data from a single taxi zone, i.e., Brooklyn Heights. The CCF plot for TNC usage and Taxi usage in Figure 12.2 shows significant negative correlation up to nearly 20 lags, with some evidence of nonstationary behavior.

```
ride.brooklyn <-
  read.csv("ride.brooklyn.csv",
          header = TRUE,
          stringsAsFactors = FALSE)
n <- n_distinct(ride.brooklyn$date)
n.hold <- 5
n.train <- n - n.hold
train.df <- ride.brooklyn[1:n.train, ]
test.df <- tail(ride.brooklyn, n.hold)
```

We construct a vectorized form of the bivariate response time series. $N = 2n$ represents the length of the 2-dimensional time series (i.e., $q = 2$) stacked one below the other, and the response vector $\boldsymbol{y}$ is the stacked time series consisting of scaled TNC and Taxi usage.

```
N <- 2 * n
y <-
  c(c(train.df$tnc, rep(NA, n.hold)), c(train.df$taxi, rep(NA, n.hold)))
```

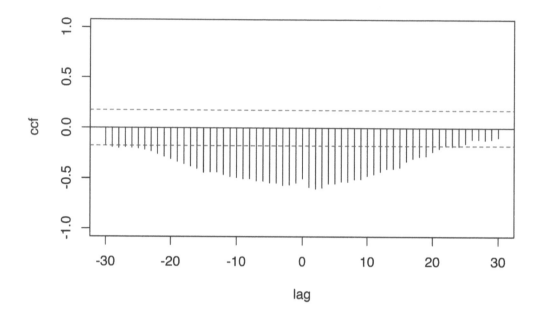

FIGURE 12.2: Sample CCF plot of TNC usage and Taxi usage.

We include Subway usage, holiday, and precipitation (these were described in Chapter 6) as exogenous variables with fixed effects (similar to the best fitting models in Chapter 6). Since $q = 2$, we construct two new variables for each predictor, one for each response. For example, to understand the impact of subway on TNC and Taxi usage, the new set of predictors is *subway.tnc* and *subway.taxi*. In *subway.tnc*, the last n values are NA's since the last n values in y correspond to data for scaled Taxi usage. Likewise, the first n values of *subway.taxi* are set as NA. In a similar fashion, we create indexes for the level, holidays, and precipitation.

```
alpha.tnc <- c(rep(1,n), rep(NA,n))
alpha.taxi <- c(rep(NA,n), rep(1,n))
subway.tnc <- c(ride.brooklyn$subway, rep(NA, n))
subway.taxi <- c(rep(NA, n), ride.brooklyn$subway)
holidays.tnc <- c(ride.brooklyn$holidays, rep(NA, n))
holidays.taxi <- c(rep(NA, n), ride.brooklyn$holidays)
precip.tnc <- c(ride.brooklyn$precip, rep(NA, n))
precip.taxi <- c(rep(NA, n), ride.brooklyn$precip)
t.index <- 1:N
b0.tnc <- c(1:n, rep(NA, n))
b0.taxi <- c(rep(NA, n), 1:n)
```

We then create a data frame as:

```
ride.mvn.data <- cbind.data.frame(y, alpha.tnc, alpha.taxi,
                        subway.tnc, subway.taxi,
                        holidays.tnc, holidays.taxi,
                        precip.tnc, precip.taxi,
                        b0.tnc, b0.taxi,
                        t.index)
```

We fit a model to $\boldsymbol{y}_t$ (consisting of time series of TNC and Taxi usage) as a function of a dynamic trend $\boldsymbol{\beta}_{t,0}$ and static effects due to Subway usage ($Z_{t,1}$), holidays ($Z_{t,2}$), and precipitation ($Z_{t,3}$). The observation equations are

$$Y_{1,t} = \alpha_1 + \beta_{1,t,0} + \beta_{1,1}Z_{t,1} + \beta_{1,2}Z_{t,2} + \beta_{1,3}Z_{t,3} + v_{1,t},$$
$$Y_{2,t} = \alpha_2 + \beta_{2,t,0} + \beta_{2,1}Z_{t,1} + \beta_{2,2}Z_{t,2} + \beta_{2,3}Z_{t,3} + v_{2,t}, \qquad (12.4)$$

where the observation error vector $\boldsymbol{v}_t \sim N(\boldsymbol{0}, \boldsymbol{V})$ with $\boldsymbol{V}$ being a 2×2 p.d. covariance matrix. The state equation corresponding to the latent vector $\boldsymbol{\beta}_{t,0} = (\beta_{1,t,0}, \beta_{2,t,0})'$ is

$$\boldsymbol{\beta}_{t,0} = \boldsymbol{\Phi}\boldsymbol{\beta}_{t-1,0} + \boldsymbol{w}_t, \qquad (12.5)$$

where $\boldsymbol{w}_t \sim N(\boldsymbol{0}, \boldsymbol{W})$, and $\boldsymbol{W}$ and $\boldsymbol{\Phi}$ are diagonal. This corresponds to $q = 2$ independent AR(1) processes.

We indicate the bivariate Gaussian assumption on $\boldsymbol{V}$ by setting `model="iid2d"` in the formula chunk for *t.index*. We also assume a Gaussian prior with high precision through the argument `control.family`, which is necessary in R-INLA for fitting MV models.

```
formula.ride.mvn.model1 <-  y ~ -1 + alpha.tnc + alpha.taxi +
   subway.tnc + subway.taxi +
   holidays.tnc + holidays.taxi +
   precip.tnc + precip.taxi +
  f(t.index, model="iid2d", n=N) +
  f(b0.tnc,model="ar1") +
  f(b0.taxi,model="ar1")
```

```
ride.mvn.model1 <-
  inla(
    formula.ride.mvn.model1,
    family = c("gaussian"),
    data = ride.mvn.data,
    control.compute = list(dic = TRUE,
                            config = TRUE),
    control.predictor = list(compute = TRUE),
    control.family =
      list(hyper = list(prec = list(
        initial = 10,
        fixed = TRUE
      )))),
  )
```

```
format.inla.out(ride.mvn.model1$summary.hyperpar[c(1,2)])
##    name                              mean      sd
## 1 Precision for t.index component 1  91.196   12.231
```

```
## 2 Precision for t.index component 2   83.942   11.335
## 3 Rho1:2 for t.index                   0.088    0.093
## 4 Precision for b0.tnc                 12.726    7.347
## 5 Rho for b0.tnc                        0.994    0.003
## 6 Precision for b0.taxi               263.960  164.675
## 7 Rho for b0.taxi                       0.956    0.029
```

In the hyperparameter summary, `Precision for t.index` (component 1) estimates the precision for TNC, $1/V_{11}$, `Precision for t.index` (component 2) estimates the precision for Taxi, $1/V_{22}$, and `Rho1:2 for t.index` estimates the correlation between R_{12}, which is not significant. The estimates of the precisions for the state errors are given by `Precision for b0.tnc` and `Precision for b0.taxi`, while the estimates for ϕ_1 and ϕ_2 are given by `Rho for b0.tnc` and `Rho for b0.taxi`.

Fitted values and forecasts in multivariate DLMs

The steps used to obtain fitted values in the multivariate framework are different from what we presented in Chapter 3 for the univariate case. We can adopt two different coding approaches to obtain fitted values in the multivariate case. The first approach uses `inla.posterior.samples()` to construct samples of the predictive distributions of the responses. The second approach uses `inla.make.lincomb()` and then fits the model.

Approach 1.

Recall that `inla.posterior.samples()` provides M (a user chosen number; here, $M = 1000$) posterior samples for the model parameters and hyperparameters. For each $m = 1, \ldots, M$, we can use these posterior values to *simulate fitted responses* from the model structure in (12.4) and (12.5). For each of the $q = 2$ components, the sample average of these M fitted response values for each time t constitutes a fitted time series, which can then be plotted against the observed TNC and Taxi usages, as shown in Figure 12.3.

```
post.sample <-
  inla.posterior.sample(1000, ride.mvn.model1)
```

We source the function below which computes the fitted values based on the model formula given the posterior samples.

```
fit.post.sample <- function(x) {
  sigma2.v1 <- 1 / x$hyperpar[1]
  sigma2.v2 <- 1 / x$hyperpar[2]
  rho.v1v2 <- x$hyperpar[3]
  cov.v1v2 <- rho.v1v2 * sqrt(sigma2.v1) * sqrt(sigma2.v2)
  sigma.v <-
    matrix(c(sigma2.v1, cov.v1v2, cov.v1v2, sigma2.v2), nrow = 2)
  V <- rmvnorm(n, mean = c(0, 0), sigma = sigma.v)
  V1 <- V[, 1]
  V2 <- V[, 2]
  fit.tnc <- x$latent[grep("alpha.tnc", rownames(x$latent))] +
    x$latent[grep("b0.tnc", rownames(x$latent))] +
    x$latent[grep("subway.tnc", rownames(x$latent))] *
    ride.brooklyn$subway +
    x$latent[grep("holidays.tnc", rownames(x$latent))] *
    ride.brooklyn$holidays +
    x$latent[grep("precip.tnc", rownames(x$latent))] *
```

```
      ride.brooklyn$precip +
      V1
  fit.taxi <- x$latent[grep("alpha.taxi", rownames(x$latent))] +
      x$latent[grep("b0.taxi", rownames(x$latent))] +
      x$latent[grep("subway.taxi", rownames(x$latent))] *
      ride.brooklyn$subway +
      x$latent[grep("holidays.taxi", rownames(x$latent))] *
      ride.brooklyn$holidays +
      x$latent[grep("precip.taxi", rownames(x$latent))] *
      ride.brooklyn$precip +
      V2
  return(list(fit.tnc = fit.tnc, fit.taxi = fit.taxi))
}
```

The fitted values for TNC usage and Taxi usage can be obtained for each set of generated parameters, and the averages over $M = 1000$ samples are saved as *tnc.fit* and *taxi.fit*.

```
fits <- post.sample %>%
  lapply(function(x)
    fit.post.sample(x))
tnc.fit <- fits %>%
  sapply(function(x)
    x$fit.tnc) %>%
  rowMeans()
taxi.fit <- fits %>%
  sapply(function(x)
    x$fit.taxi) %>%
  rowMeans()
```

Figure 12.3 shows the fitted versus observed values for TNC usage (top plot) and Taxi usage (bottom plot). The model fit is close for most of the training period. While the forecasts for Taxi usage in the test period are accurate, the model seems to under predict actual TNC usage which appears to have an upswing for those time periods.

Approach 2.

We set up and use linear combinations of the latent random field (for details, see Martins et al. (2013)). Specifically, for model fitting and prediction, we use inla.make.lincomb(), which creates a list for every t with *idx* and *weight*, where *weight* is the value of the predictor corresponding to the time index *idx*. The setup for TNC is shown below.

```
lc1 <- c()
tnc.dat.lc <- ride.brooklyn %>%
  select(subway, holidays:precip)
for(i in 1:n){
  idx0 <- idx1 <- idx2 <- idx3 <-  rep(0,n)
  idx0[i] <-  1
  idx1[i] <- tnc.dat.lc[i,1]
  idx2[i] <- tnc.dat.lc[i,2]
  idx3[i] <- tnc.dat.lc[i,3]
  lcc <-  inla.make.lincomb(alpha.tnc=1, b0.tnc = idx0, subway.tnc=idx1[i],
                            holidays.tnc = idx2[i], precip.tnc = idx3[i])
  for (k in 1:length(lcc[[1]])) {
```

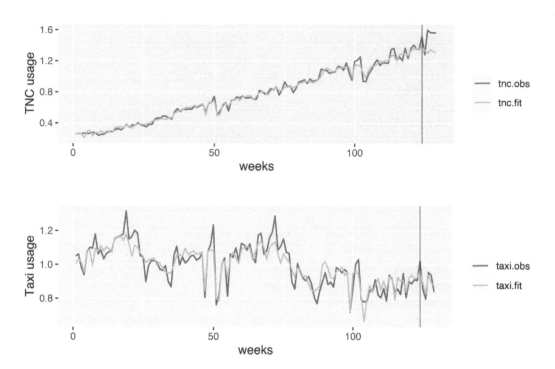

FIGURE 12.3: Observed (red) versus fitted (blue) TNC (top panel) and Taxi usage (bottom panel). Fits obtained using `inla.posterior.sample()`. The black vertical line divides the data into training and test portions.

```
    if(length(lcc[[1]][[k]][[1]]$weight)==0){
      lcc[[1]][[k]][[1]]$idx <- i
      lcc[[1]][[k]][[1]]$weight <- 0
    }
  }
  names(lcc) <- paste0("tnc", i)
  lc1 <- c(lc1, lcc)
}
```

The setup for Taxi usage is similar.

```
lc2 <- c()
taxi.dat.lc <- ride.brooklyn %>%
  select(subway, holidays:precip)
for (i in 1:n) {
  idx0 <- idx1 <- idx2 <- idx3 <- rep(0, n)
  idx0[i] <- 1
  idx1[i] <- taxi.dat.lc[i, 1]
  idx2[i] <- taxi.dat.lc[i, 2]
  idx3[i] <- taxi.dat.lc[i, 3]
  lcc <-
    inla.make.lincomb(
      alpha.taxi = 1,
```

```
      b0.taxi = idx0,
      subway.taxi = idx1[i],
      holidays.taxi = idx2[i],
      precip.taxi = idx3[i]
    )
  for (k in 1:length(lcc[[1]])) {
    if (length(lcc[[1]][[k]][[1]]$weight) == 0) {
      lcc[[1]][[k]][[1]]$idx <- i
      lcc[[1]][[k]][[1]]$weight <- 0
    }
  }
  names(lcc) <- paste0("taxi", i)
  lc2 <- c(lc2, lcc)
}
```

We then combine the linear combinations into *lc* as follows.

```
lc <- c(lc1, lc2)
```

The list *lcc* contains the index number (time), the values for fixed predictors, and latent intercepts $\beta_{1,t,0}$ and $\beta_{2,t,0}$. For example, the components of lc1[[5]] will show values of the predictors at time $t = 5$:

```
lc1[[5]]
```

We include the list of linear combinations created in our model fitting using the argument `lincomb = lc` in the `inla()` call. This enables us to obtain posterior marginals of the linear combinations, as shown below. Also note that the model output will not include a "Precision for Gaussian observation" term, due to our assumption of a large and fixed precision.

```
ride.mvn.model1 <-
  inla(
    formula.ride.mvn.model1,
    family = c("gaussian"),
    data = ride.mvn.data,
    control.compute = list(dic = TRUE, config = TRUE),
    control.predictor = list(compute = TRUE),
    control.family =
      list(hyper = list(prec = list(
        initial = 10,
        fixed = TRUE
      ))),
    lincomb = lc
  )
format.inla.out(ride.mvn.model1$summary.hyperpar[c(1, 2)])
## name                                  mean      sd
## 1 Precision for t.index component 1   91.197    12.231
## 2 Precision for t.index component 2   83.943    11.335
## 3 Rho1:2 for t.index                  0.088     0.093
## 4 Precision for b0.tnc                12.654    7.318
## 5 Rho for b0.tnc                      0.995     0.003
## 6 Precision for b0.taxi               262.717   163.221
## 7 Rho for b0.taxi                     0.956     0.029
```

Plots of fitted versus observed usage for TNC and Taxi usage are shown in Figure 12.4.

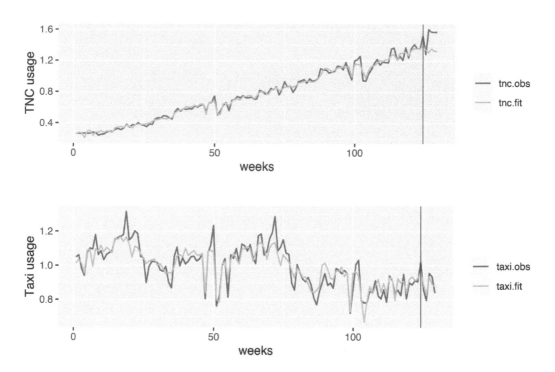

FIGURE 12.4: Observed (red) versus fitted (blue) TNC usage (top panel) and Taxi usage (bottom panel). Fits obtained using `inla.make.lincomb()`. The black vertical line divides the data into training and test portions.

We end this section with a few useful remarks.

Remark 1. If we wish to modify the inverse Wishart prior specification, we can use the code given below.

```
formula.ride.mvn.model1 <-  y ~ -1 +
  f(t.index,
    model = "iid2d",
    n = N,
    hyper = list(theta1 = list(param = c(4, 0.01, 0.01, 0))))) + ...
```

We see that differences due to the different prior specifications are minimal for large n.

Remark 2. If we wish to include the effects of the exogenous predictors as dynamic *random effects*, rather than fixed effects as shown above, the following code is useful. Setting up the linear combinations is similar to the fixed effects case; the only change is that we replace the name of the predictor by its index. For example, we replace *subway.tnc* (which contains data on Subway usage for the first n rows and NA's for the next n rows) with *id.sub.tnc* (which is a vector with $1 : n$ followed by n NA's). The variables *subway.taxi* and *id.sub.taxi* will have the Subway usage data or $1 : n$ switched with the NA's.

```
t.index <- 1:N
alpha.tnc <- c(rep(1,n), rep(NA,n))
alpha.taxi <- c(rep(NA,n), rep(1,n))
```

```
b0.tnc <- id.sub.tnc <-
  id.holidays.tnc <- id.precip.tnc <- c(1:n, rep(NA, n))
b0.taxi <- id.sub.taxi <- id.events.taxi <-
  id.holidays.taxi <- id.precip.taxi <- c(rep(NA, n), 1:n)
ride.mvn.data <- cbind.data.frame(
  y,alpha.tnc,alpha.taxi,subway.tnc,subway.taxi,holidays.tnc,
  holidays.taxi,precip.tnc,precip.taxi,
  b0.tnc,b0.taxi,id.sub.tnc,
  id.sub.taxi,id.holidays.tnc,id.holidays.taxi,
  id.precip.tnc,id.precip.taxi,t.index
)
```

In the formula, we then replace the fixed effects by suitable random effect specifications as shown below.

```
formula.ride.mvn.model1 <-  y ~ -1 + alpha.tnc + alpha.taxi +
  f(t.index, model = "iid2d", n = N) +
  f(b0.tnc, model = "ar1", constr = FALSE) +
  f(b0.taxi, model = "ar1", constr = FALSE) +
  f(id.sub.tnc, subway.tnc, model = "ar1", constr = FALSE) +
  f(id.sub.taxi, subway.taxi, model = "ar1", constr = FALSE) +
  f(id.holidays.tnc,
    holidays.tnc,
    model = "ar1",
    constr = FALSE) +
  f(id.holidays.taxi,
    holidays.taxi,
    model = "ar1",
    constr = FALSE) +
  f(id.precip.tnc,
    precip.tnc,
    model = "ar1",
    constr = FALSE) +
  f(id.precip.taxi,
    precip.taxi,
    model = "ar1",
    constr = FALSE)
```

Corresponding linear combinations can then be similarly constructed, as shown below for TNC usage. The construction for Taxi usage will be similar.

```
lc1 <- c()
tnc.dat.lc <- ride.brooklyn %>%
  select(subway, holidays:precip)
for(i in 1:n){
  idx0 <- idx1 <- idx2 <- idx3 <-  rep(0,n)
  idx0[i] <-  1
  idx1[i] <- tnc.dat.lc[i,1]
  idx2[i] <- tnc.dat.lc[i,2]
  idx3[i] <- tnc.dat.lc[i,3]
  lcc <-  inla.make.lincomb(b0.tnc=idx0, id.sub.tnc=idx1,
                     id.holidays.tnc = idx2, id.precip.tnc = idx3)
```

```
  for (k in 1:length(lcc[[1]])) {
    if(length(lcc[[1]][[k]][[1]]$weight)==0){
      lcc[[1]][[k]][[1]]$idx <- i
      lcc[[1]][[k]][[1]]$weight <- 0
    }
  }
  names(lcc) <-  paste0("tnc", i)
  lc1 <-  c(lc1, lcc)
}
```

Remark 3. Note that the current version of the function `inla.make.lincomb()` gives an error when a predictor value is 0 (for example, in holidays or precipitation), the reason being that a *weight=0* corresponding to this 0 value has now been coded as NA in R-INLA.[1] To circumvent this, we include a workaround by the following lines of code which replace *weight = NA* by *weight=0*:

```
for (k in 1:length(lcc[[1]])) {
    if(length(lcc[[1]][[k]][[1]]$weight)==0){
      lcc[[1]][[k]][[1]]$idx <- i
      lcc[[1]][[k]][[1]]$weight <- 0
    }
  }
```

12.3   Model with equicorrelated w_t and diagonal Φ

Let Φ be diagonal, while $Cov(\boldsymbol{w}_t) = \boldsymbol{W} = \sigma_w^2[(1-\rho_w)\boldsymbol{I} + \rho_w\boldsymbol{J}]$, where $\boldsymbol{I}$ is the $p \times p$ identity matrix and $\boldsymbol{J}$ is the $p \times p$ matrix whose entries are all 1. That is, components of the state error vector $\boldsymbol{w}_t$ are correlated, such that the correlation between any pair of components is the same $|\rho_w| < 1$ (intra-class correlation or equi-correlation structure). We model simulated trivariate Gaussian time series with this structure using the *group* option.

12.3.1   Example: Simulated trivariate series

We simulate a trivariate series $\boldsymbol{y}_t$. The observation error covariance $\boldsymbol{V}$ is set up below.

```
V.11 <- 1.0
V.22 <- 1.0
V.33 <- 1.0
R.12 <- 0.7
R.13 <- 0.4
R.23 <- 0.25
V.12 <- R.12 * sqrt(V.11) * sqrt(V.22)
V.13 <- R.13 * sqrt(V.11) * sqrt(V.33)
V.23 <- R.23 * sqrt(V.22) * sqrt(V.33)
V <- matrix(c(V.11, V.12, V.13, V.12, V.22, V.23,
              V.13, V.23, V.33), nrow = 3)
```

---

[1] private communication with Håvard Rue (https://www.r-inla.org/contact-us)

The state error covariance $\boldsymbol{W}$ has the equicorrelation structure.

```
sig2.w <- 0.1
rho.w <- 0.75
W.11 <- sig2.w
W.22 <- sig2.w
W.33 <- sig2.w
W.12 <- sig2.w*rho.w
W.13 <- sig2.w*rho.w
W.23 <- sig2.w*rho.w
W <- matrix(c(W.11, W.12, W.13, W.12, W.22, W.23,
              W.13, W.23, W.33), nrow = 3)
```

We assume that each AR(1) coefficient has the same value ϕ. We then set up the arrays $\boldsymbol{I}_3$ and $\mathrm{diag}(\phi_1, \phi_2, \phi_3)$ to be fed as parameters in the `marima.sim()` function in the R package *marima*. Note that the ϕ value in this package corresponds to the AR polynomial $(1 + \phi B)$ and not $(1 - \phi B)$, which is the usual convention.

```
phi.1 <- -0.5
phi.2 <- -0.5
phi.3 <- -0.5
# Set up Phi as be a 3-dim array
Phi <- array(0, dim = c(3, 3, 2))
Phi[1,1,1] <- 1.0
Phi[2,2,1] <- 1.0
Phi[3,3,1] <- 1.0
Phi[1,1,2] <- phi.1
Phi[2,2,2] <- phi.2
Phi[3,3,2] <- phi.3
```

We simulate the states from VAR(1) processes using the *marima* package.

```
n <- n.sim <- 1000
q <- 3
n.start <- 0
avg <- rep(0, 3)
X <-
  marima.sim(
    3,
    ar.model = Phi,
    ar.dif = NULL,
    ma.model = NULL,
    averages = rep(0, 3),
    resid.cov = W,
    seed = 1234,
    nstart = n.start,
    nsim = n.sim
  )
x1 <- X[, 1]
x2 <- X[, 2]
x3 <- X[, 3]
```

Figure 12.5 shows the ACF plots for the three state vector components, with $\phi = -0.5$.

```
par(mfrow=c(1,3))
acf(x1, main ="", sub="ACF of x1")
acf(x2, main ="", sub="ACF of x2")
acf(x3, main ="", sub="ACF of x3")
```

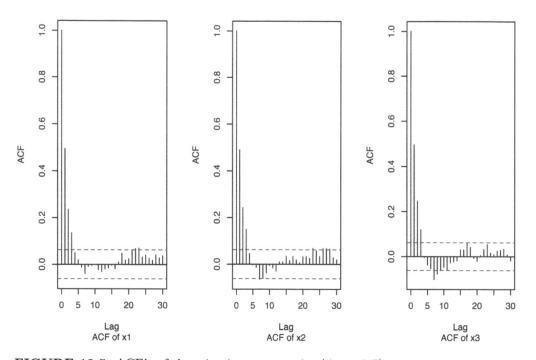

FIGURE 12.5: ACF's of the trivariate state series ($\phi = -0.5$).

We next simulate observation errors from a $N_3(\mathbf{0}, \boldsymbol{V})$ distribution, and add these to the states to get the response time series $\boldsymbol{y}_t$. Figure 12.6 shows the CCF of $\boldsymbol{x}_t$ (top row), and the CCF of $\boldsymbol{y}_t$ (bottom row).

```
# Generate observation errors
set.seed(123457)
err.v <- rmvnorm(n.sim, mean = c(0, 0, 0), sigma = V)
err.v1 <- err.v[, 1]
err.v2 <- err.v[, 2]
err.v3 <- err.v[, 3]
# Add observation errors to states to get responses
y1 <- x1 + err.v1
y2 <- x2 + err.v2
y3 <- x3 + err.v3
y <- c(y1, y2, y3)

par(mfrow=c(2,3))
ccf(x1, x2, ylab = "CCF", main ="", sub="CCF of (x1,x2)")
ccf(x1, x3, ylab = "CCF", main ="", sub="CCF of (x1,x3)")
ccf(x2, x3, ylab = "CCF", main ="", sub="CCF of (x2,x3)")
```

```
ccf(y1,y2, main="", sub="CCF of (y1,y2)")
ccf(y1,y3, main="", sub="CCF of (y1,y3)")
ccf(y2,y3, main="", sub="CCF of (y2,y3)")
```

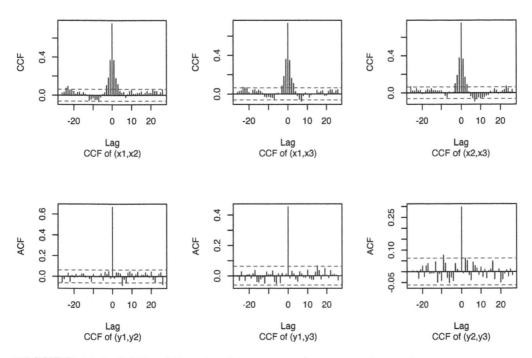

FIGURE 12.6: CCF's of the trivariate state and response time series.

The formula and model fit follow. Note that id.b0.g <- rep(1:3, each = n) allows us to incorporate a grouping that reflects the equicorrelated state errors and the same ϕ in the the diagonal of $\boldsymbol{\Phi}$ in the VAR(1) model for the state vector (dynamic intercept). This is used in the line f(id.b0, model="ar1", constr = FALSE, group = id.b0.g). As in Section 12.2, f(id.V, model="iid3d", n=N) allows a trivariate inverse Wishart specification for V.

```
m <- n-1
N <- 3*n
id.V <- 1:N
id.b0 <- c(1:n, 1:n, 1:n)
id.b0.g <- rep(1:3, each = n)
triv.dat <- cbind.data.frame(y, id.b0, id.b0.g, id.V)
formula.triv <-  y ~ 1 +
  f(id.V, model="iid3d", n=N) +
  f(id.b0, model="ar1", constr = FALSE, group = id.b0.g)

result.triv <-
  inla(
    formula.triv,
    family = c("gaussian"),
```

```
    data = triv.dat,
    control.inla = list(h = 1e-5, tolerance = 1e-3),
    control.compute = list(dic = TRUE, config = TRUE),
    control.predictor = list(compute = TRUE),
    control.family = list(hyper = list(prec = list(
      initial = 12, fixed = TRUE
    )))
  )
result.triv.summary <- summary(result.triv)

format.inla.out(result.triv.summary$hyperpar[c(1,2)])
## name                                   mean      sd
## 1 Precision for id.V component 1       0.958 2.700e-02
## 2 Precision for id.V component 2       0.951 2.700e-02
## 3 Precision for id.V component 3       0.860 2.800e-02
## 4 Rho1:2 for id.V                      0.668 1.300e-02
## 5 Rho1:3 for id.V                      0.453 1.900e-02
## 6 Rho2:3 for id.V                      0.299 2.000e-02
## 7 Precision for id.b0               1304.921 5.537e+16
## 8 Rho for id.b0                       -0.520 2.180e-01
## 9 GroupRho for id.b0                   0.869 6.600e-02
```

The first three rows in the output provide posterior summaries for the precisions of the components of v_t, whose true values are 1.0. The next three rows are posterior summaries of the correlations between the components of v_t, whose true values are $R_{12} = 0.7, R_{13} = 0.4$, and $R_{23} = 0.25$. The value Rho for id.b0 estimates the common ϕ, with true value -0.5. The term GroupRho for id.b0 estimates the equicorrelation coefficient ρ_w, whose true value is 0.75. We can recover σ_w^2 by dividing *Precision for id.b0* by $1 - \phi^2$. Again, to obtain fitted values and predictions, we can either use inla.posterior.samples() or inla.make.lincomb() (see the detailed discussion in Section 12.2).

12.4   Fitting multivariate models using rgeneric

In vector time series modeling, we often need to fit models with a full (non-diagonal) Φ matrix, and/or an unstructured state error covariance matrix W. To the best of our knowledge, no R-INLA template currently exists for fitting such a model. Instead, we implement this by using the rgeneric() function. Gómez-Rubio (2020) used the rgeneric() function to fit a conditionally autoregressive (CAR) specification to latent spatial random effects, while Palmí-Perales et al. (2021) discussed multivariate spatial models for lattice data. For more details, see the example discussed in vignette("rgeneric", package = INLA).

Here, we illustrate the use of the rgeneric() function for fitting a VAR(1) state equation (see Dutta et al. (2022) for more details). In using rgeneric(), the user directly computes the likelihood and priors that are required for the model specification of interest. The function inla.rgeneric.define() defines the different latent effects, and takes as arguments functions such as graph(), Q(), etc., along with other arguments that are required to evaluate the different functions.

12.4.1 Example: Simulated bivariate VAR(1) series

We simulate bivariate time series from a model whose observation and state equations are given below:

$$
\begin{aligned}
\boldsymbol{y}_t &= \boldsymbol{x}_t + \boldsymbol{v}_t, \\
\boldsymbol{x}_t &= \boldsymbol{\Phi}\boldsymbol{x}_{t-1} + \boldsymbol{w}_t,
\end{aligned}
\tag{12.6}
$$

where $\boldsymbol{v}_t \sim N(\boldsymbol{0}, \boldsymbol{V})$ with $\boldsymbol{V} = \begin{pmatrix} 1/\tau_{11} & R_{12}/\sqrt{\tau_{11}\tau_{22}} \\ R_{12}/\sqrt{\tau_{11}\tau_{22}} & 1/\tau_{22} \end{pmatrix}$, τ_{ii} is the marginal precision of the ith component of $\boldsymbol{V}$, $i = 1, 2$, and R_{12} is the correlation between the two components of $\boldsymbol{V}$. Here, $\boldsymbol{\Phi} = \begin{pmatrix} \phi_{11} & \phi_{12} \\ \phi_{21} & \phi_{22} \end{pmatrix}$, and $\boldsymbol{W} = \mathrm{diag}(1/\tau_{w1}, 1/\tau_{w2})$, τ_{wi} being the marginal precision of the ith component of $\boldsymbol{w}_t$, $i = 1, 2$. Let $\boldsymbol{\Theta} = (\phi_{11}, \phi_{12}, \phi_{21}, \phi_{22}, \tau_{w1}, \tau_{w2}, \tau_{11}, \tau_{22}, R_{12})'$. The data is simulated using the code below and these true parameter values:

$$
\boldsymbol{\Phi} = \begin{pmatrix} 0.7 & 0.1 \\ 0.3 & 0.2 \end{pmatrix}, \ \boldsymbol{V} = \begin{pmatrix} 1/2 & 0.55/\sqrt{2 \times 3} \\ 0.55/\sqrt{2 \times 3} & 1/3 \end{pmatrix}, \text{ and } \boldsymbol{W} = \begin{pmatrix} 1/2 & 0 \\ 0 & 1/2 \end{pmatrix}.
$$

```
set.seed(123456)
p <- 1
k <- 2
n <- 300
phi.mat <- matrix(c(0.7,0.3,0.1,0.2),nrow=2,byrow=TRUE)
# Construct the W matrix
init.val <- matrix(c(0, 0))
rho.w1w2 = 0
sigma2.w1 <- 1 / 2
sigma2.w2 <- 1 / 2
cov.w1w2 <- rho.w1w2 * sqrt(sigma2.w1 * sigma2.w2)
sigma.mat.w <-
  matrix(c(sigma2.w1, cov.w1w2, cov.w1w2, sigma2.w2),
         nrow = 2,
         byrow = T)

sample.list <- list()
sample.list[[1]] <- init.val
for (i in 2:n) {
  sample.list[[i]] <-
    phi.mat %*% sample.list[[i - 1]] + matrix(MASS::mvrnorm(
      n = 1,
      mu = c(0, 0)
      ,
      Sigma = sigma.mat.w
    ))
}
simulated_var <- do.call(cbind, sample.list)
var.ts <- t(simulated_var)
colnames(var.ts) <- c("y1", "y2")
x.t <- vec(var.ts)
```

We set up the true $\boldsymbol{V}$ matrix and simulate $\boldsymbol{y}_t$.

```
rho.v1v2 = 0.55
sigma2.v1 <- 1 / 2
sigma2.v2 <- 1 / 3
cov.v1v2 <- rho.v1v2 * sqrt(sigma2.v1 * sigma2.v2)
sigma.mat.v <-
  matrix(c(sigma2.v1, cov.v1v2, cov.v1v2, sigma2.v2),
         nrow = 2,
         byrow = T)
v.t <- vec(mvrnorm(n, mu = c(0, 0), Sigma = sigma.mat.v))
y.t <- x.t + v.t
```

We next set up indexes and the data frame.

```
inla.dat <- data.frame(y = y.t, iid2d.id = 1:length(y.t))
odd.id <- seq(1, nrow(inla.dat), by = 2)
even.id <- seq(2, nrow(inla.dat), by = 2)
var.id <- c(odd.id, even.id)
inla.dat$var.id <- var.id
```

We source the function `inla.rgeneric.VAR_1()`, which is included in the list of custom functions, and is shown in Chapter 14. Brief descriptions of the functions used within `inla.rgeneric.VAR_1()` are given below. For details on the formulas, see Lütkepohl (2013), Rue and Held (2005), and Chapter 12 – Appendix.

1. `graph()` defines which entries of a user defined precision matrix `Q()` are non-zero:

```
graph = function() {
    return (inla.as.sparse(Q()))
  }
```

2. The precision matrix $Q(\theta)$ is computed in sparse form by using the function `Q()`, with **param** denoting the model parameters:

```
Q <- function() {
  param <- interpret.theta()
  phi.mat <- matrix(param$phi.vec, nrow = k)
  sigma.w.inv <- param$PREC
  A <- t(phi.mat) %*% sigma.w.inv %*% phi.mat
  B <- -t(phi.mat) %*% sigma.w.inv
  C <- sigma.w.inv
  # Construct mid-block
  zero.mat <- matrix(0, nrow = 2, ncol = 2 * n)
  # Define the matrix block
  mat <- cbind(t(B), A + C, B)
  # Initializing column id and matrix list
  col.id <- list()
  mat.list <- list()
  col.id[[1]] <- 1:(3 * k)
  mat.list[[1]] <- zero.mat
  mat.list[[1]][, col.id[[1]]] <- mat
  for (id in 2:(n - 2)) {
```

```
      start.id <- col.id[[id - 1]][1] + k
      end.d <-  start.id + (3 * k - 1)
      col.id[[id]] <- start.id:end.d
      mat.list[[id]] <- zero.mat
      mat.list[[id]][, col.id[[id]]] <- mat
  }
  mid.block <- do.call(rbind, mat.list)
  tau.val <- 0.1
  diffuse.prec <- tau.val * diag(1, k)
  # Construct first and last row blocks and then join with mid block
  first.row.block <-
      cbind(A + diffuse.prec, B, matrix(0, nrow = k, ncol = (k * n - k ^ 2)))
  last.row.block <- cbind(matrix(0, nrow = k, ncol = k * n - k ^ 2), t(B),C)
  toep.block.mat <-
      rbind(first.row.block, mid.block, last.row.block)
  # Changing to a sparse Matrix
  prec.mat <- Matrix(toep.block.mat, sparse = TRUE)
  return(prec.mat)
}
```

3. The function mu() is a zero vector:

```
mu <- function() {
      return(numeric(0))
    }
```

4. The function initial() contains initial values of the parameter Θ expressed in R-INLA's internal scale:

```
initial = function() {
  phi.init <- c(0.1, 0.1, 0.1, 0.1)
  prec.init <- rep(1, k)
  init <- c(phi.init, prec.init)
  return (init)
}
```

5. log.norm.constant() denotes the logarithm of the normalizing constant from a multivariate Gaussian distribution. If we assign numeric(0), R-INLA will compute this constant.

```
log.norm.const <- function() {
  Q <- Q()
  log.det.val <-
    Matrix::determinant(Q, logarithm = TRUE)$modulus
  val <- (-k * n / 2) * log(2 * pi) + 0.5 * log.det.val
  return (val)
}
```

6. The function log.prior() specifies the priors for the original model parameters on the log scale. Note that although R-INLA works with internal variables (see

Chapter 3), the log prior is set on the original parameters, such as $\tau_{11}, \tau_{22}, R_{12}$, etc. We also include a term for the logarithm of the Jacobian of the transformations from the original to the internal scale of the parameters. Specifically, we set up a normal prior for each of the elements of the $\mathbf{\Phi}$ matrix, and log gamma priors for each of the precisions from the $\mathbf{W}$ matrix.

```r
log.prior <- function() {
  param <- interpret.theta()
  pars <- param$param
  k <- k
  total.par <- param$n.phi + param$n.prec
  # Normal prior for phi's
  theta.phi <- theta[1:param$n.phi]
  phi.prior <-
    sum(sapply(theta.phi, function(x)
      dnorm(
        x, mean = 0, sd = 1, log = TRUE
      )))
  theta.prec <-
    theta[(param$n.phi + 1):(param$n.phi + param$n.prec)]
  prec.prior <-
    sum(sapply(theta.prec, function(x)
      dgamma(
        x,
        shape = 1,
        scale = 1,
        log = TRUE
      )))
  prec.jacob <- sum(theta.prec)
  prior.val <- phi.prior + prec.prior + prec.jacob
  return (prior.val)
}
```

The different functions described above are then pulled into the `inla.rgeneric.VAR_1()` function. The complete set of functions can be accessed from `inla.rgeneric.VAR_1()` in Section 14.3.1. We can also obtain it from the GitHub link for the book. If we use R version 4.0 or higher, R-INLA gives a small fix that we must add to our code. These are also shown in the code below.

```r
inla.rgeneric.VAR_1 <-
  function(cmd = c("graph",
                   "Q",
                   "mu",
                   "initial",
                   "log.norm.const",
                   "log.prior",
                   "quit"),
           theta = NULL)
  {
    envir = parent.env(environment())
    graph = function() {
      ...
```

```
  }
  Q = function() {
    ...
  }
  mu = function() {
    ...
  }
  log.norm.const = function() {
    ...
  }
  log.prior = function() {
    ...
  }
  initial = function() {
    ...
  }
  quit = function() {
    ...
  }
# Fix for R version higher than 4.0
  if (as.integer(R.version$major) > 3) {
    if (!length(theta))
      theta <- initial()
  } else {
    if (is.null(theta)) {
      theta <- initial()
    }
  }
  val <- do.call(match.arg(cmd), args = list())
  return (val)
```

The function `interpret.theta()` enables us to obtain the list of original parameters, along with the items used for the setup, such as the number of elements in $\mathbf{\Phi}$, number of precision parameters in $\mathbf{W}$, etc.

```
interpret.theta <- function(theta) {
  n.phi <- k * k * p
  n.prec <- k
  n.tot <- n.phi + n.prec
  phi.VAR <- sapply(theta[as.integer(1:n.phi)], function(x) {
    x
  })
  # W matrix precisions
  wprec <-
    sapply(theta[as.integer((n.phi + 1):(n.phi + n.prec))], function(x) {
      exp(x)
    })
  param <- c(phi.VAR, wprec)
  W <- diag(1, n.prec)
  st.dev <- 1 / sqrt(wprec)
  st.dev.mat <-
```

```
      matrix(st.dev, ncol = 1) %*% matrix(st.dev, nrow = 1)
    W <- W * st.dev.mat
    PREC <- solve(W)
    return(list(
      param = param,
      VACOV = W,
      PREC = PREC,
      phi.vec = c(phi.VAR),
      n.phi = n.phi,
      n.prec = n.prec
    ))
}

model.define <-
  inla.rgeneric.define(inla.rgeneric.VAR_1,
                       p = 1,
                       k = 2,
                       n = n)

result.rgeneric <- inla(
  formula.rgeneric,
  family = c("gaussian"),
  data = inla.dat,
  control.compute = list(dic = TRUE, config = TRUE),
  control.predictor = list(compute = TRUE),
  control.mode = list(restart = TRUE),
  control.family = list(hyper = list(prec = list(
    initial = 15,
    fixed = T
  ))),
  verbose = TRUE
)
#summary(result.rgeneric)
format.inla.out(result.rgeneric$summary.hyperpar[, c(1, 2, 4)])
```

```
##    name                                    mean  sd    0.5q
## 1 Theta1 for var.id                        0.709 0.052 0.709
## 2 Theta2 for var.id                        0.106 0.041 0.106
## 3 Theta3 for var.id                        0.340 0.092 0.339
## 4 Theta4 for var.id                        0.211 0.072 0.211
## 5 Theta5 for var.id                        0.581 0.237 0.582
## 6 Theta6 for var.id                        0.833 0.181 0.832
## 7 Precision for iid2d.id component 1       2.519 0.606 2.447
## 8 Precision for iid2d.id component 2       3.004 0.667 2.937
## 9 Rho1:2 for iid2d.id                      0.441 0.095 0.447
```

We run the multivariate DLM model and summarize the true and estimated parameters below. The estimated posterior means are close to the true parameters, even for a moderate sample size of $n = 300$.

Remark. Models with other specifications for the state vector can be fit by suitably modifying the functions shown above, particularly the functions Q(), log.norm.const(), and `log.prior()`. Since the source code is fully written in R, the computational time may be high.

Chapter 12 – Appendix

Vector AR models

Let $\boldsymbol{x}_t = (X_{t,1}, \ldots, X_{t,k})'$, with mean $E(\boldsymbol{x}_t) = \boldsymbol{\mu}$. We define the vector autoregressive model of order p, VAR(p) model, as

$$\boldsymbol{x}_t = \boldsymbol{\alpha} + \sum_{j=1}^{p} \boldsymbol{\Phi}_j \boldsymbol{x}_{t-j} + \boldsymbol{w}_t, \tag{12.7}$$

where the error vector $\boldsymbol{w}_t \sim N(\boldsymbol{0}, \boldsymbol{W})$, $\boldsymbol{W}$ being a $k \times k$ p.d. matrix, and $\boldsymbol{\Phi}_j$, $j = 1, \ldots, p$ are $k \times k$ state transition matrices. The k-dimensional intercept $\boldsymbol{\alpha}$ can be written in terms of the mean $\boldsymbol{\mu}$ and the coefficient matrices $\boldsymbol{\Phi}_j$, $j = 1, \ldots, p$ as

$$\boldsymbol{\alpha} = (\boldsymbol{I}_k - \sum_{j=1}^{p} \boldsymbol{\Phi}_j) \boldsymbol{\mu}.$$

The VAR(p) operator is defined as

$$\boldsymbol{\Phi}(B) = \boldsymbol{I}_k - \boldsymbol{\Phi}_1 B - \ldots - \boldsymbol{\Phi}_p B^p.$$

The process is stationary if the roots of the determinant of $\boldsymbol{\Phi}(z)$ are outside the unit circle, i.e., the modulus of each root is greater than 1. For $p = 1$, the vector autoregressive model of order 1, or VAR(1) process, is defined as

$$\boldsymbol{x}_t = \boldsymbol{\alpha} + \boldsymbol{\Phi} \boldsymbol{x}_{t-1} + \boldsymbol{w}_t. \tag{12.8}$$

Let $\boldsymbol{x}_t$, $t = 1, \ldots, n$ be observed realizations from the VAR(1) process with zero mean. The joint distribution of the model parameters given the data $(\boldsymbol{x}_1', \ldots, \boldsymbol{x}_n')'$ has been derived in Dutta et al. (2022), assuming that $\boldsymbol{x}_1 \sim N(\boldsymbol{0}, \frac{1}{\kappa} \boldsymbol{I}_k)$, κ being a small number. Then,

$$\log p(\boldsymbol{x}_t | \boldsymbol{x}_{t-1}) = -\frac{1}{2} \boldsymbol{w}_t' \boldsymbol{W}^{-1} \boldsymbol{w}_t + C, \tag{12.9}$$

where $\boldsymbol{w}_t = \boldsymbol{x}_t - \boldsymbol{\Phi} \boldsymbol{x}_{t-1}$, $p(.)$ denotes the p.d.f. of the state errors, and C is a constant. To illustrate, when $n = 3$, the joint distribution of $(\boldsymbol{x}_1', \boldsymbol{x}_2', \boldsymbol{x}_3')'$ becomes

$$\log p(\boldsymbol{x}_1, \boldsymbol{x}_2, \boldsymbol{x}_3) = \log p(\boldsymbol{x}_3 | \boldsymbol{x}_2) + \log p(\boldsymbol{x}_2 | \boldsymbol{x}_1) + \log p(\boldsymbol{x}_1)$$

$$= -\frac{1}{2} \big[(\boldsymbol{x}_3 - \boldsymbol{\Phi} \boldsymbol{x}_2)' \boldsymbol{W}^{-1} (\boldsymbol{x}_3 - \boldsymbol{\Phi} \boldsymbol{x}_2) + (\boldsymbol{x}_2 - \boldsymbol{\Phi} \boldsymbol{x}_1)' \boldsymbol{W}^{-1} (\boldsymbol{x}_2 - \boldsymbol{\Phi} \boldsymbol{x}_1) + \boldsymbol{x}_1' \kappa \boldsymbol{I}_3 \boldsymbol{x}_1 \big] + C^*$$

$$= -\frac{1}{2} \big[\boldsymbol{x}_3' \boldsymbol{W}^{-1} \boldsymbol{x}_3 - \boldsymbol{x}_3' \boldsymbol{W}^{-1} \boldsymbol{\Phi} \boldsymbol{x}_2 - \boldsymbol{x}_2' \boldsymbol{\Phi}' \boldsymbol{W}^{-1} \boldsymbol{x}_3 + \boldsymbol{x}_2' \boldsymbol{\Phi}' \boldsymbol{W}^{-1} \boldsymbol{\Phi} \boldsymbol{x}_2$$

$$+ \boldsymbol{x}_2' \boldsymbol{W}^{-1} \boldsymbol{x}_2 - \boldsymbol{x}_2' \boldsymbol{W}^{-1} \boldsymbol{\Phi}' \boldsymbol{x}_1 - \boldsymbol{x}_1' \boldsymbol{\Phi}' \boldsymbol{W}^{-1} \boldsymbol{x}_2 + \boldsymbol{x}_1' (\boldsymbol{\Phi}' \boldsymbol{W}^{-1} \boldsymbol{\Phi} + \kappa \boldsymbol{I}_3) \boldsymbol{x}_1 \big],$$

$$\tag{12.10}$$

where C^* is another constant. Let

$$\boldsymbol{A} = \boldsymbol{\Phi}'\boldsymbol{W}^{-1}\boldsymbol{\Phi}, \boldsymbol{B} = -\boldsymbol{\Phi}'\boldsymbol{W}^{-1}, \text{ and } \boldsymbol{C} = \boldsymbol{W}^{-1} \qquad (12.11)$$

be $k \times k$ matrices. Then, the precision matrix corresponding to the joint distribution of $(\boldsymbol{x}_1', \boldsymbol{x}_2', \boldsymbol{x}_3')'$ is a $3k \times 3k$ matrix given by

$$\begin{bmatrix} \boldsymbol{A} + \kappa \boldsymbol{I}_k & \boldsymbol{B} & \boldsymbol{O}_k \\ \boldsymbol{B}' & \boldsymbol{A} + \boldsymbol{C} & \boldsymbol{B} \\ \boldsymbol{O}_k & \boldsymbol{B}' & \boldsymbol{C}. \end{bmatrix}$$

where $\boldsymbol{O}_k$ denotes a $k \times k$ matrix with zero entries.

Using the same approach, the precision matrix in the joint distribution of $(\boldsymbol{x}_1', \ldots, \boldsymbol{x}_n')'$ is derived as

$$\begin{bmatrix} \boldsymbol{A} + \kappa \boldsymbol{I}_k & \boldsymbol{B} & \cdots & \cdots & \cdots & \cdots & \cdots & \boldsymbol{O}_k \\ \boldsymbol{B}' & \boldsymbol{A} + \boldsymbol{C} & \boldsymbol{B} & \ddots & & & & \vdots \\ \vdots & \boldsymbol{B}' & \boldsymbol{A} + \boldsymbol{C} & \boldsymbol{B} & & & & \vdots \\ \vdots & \ddots & \ddots & \ddots & \ddots & \ddots & & \vdots \\ \vdots & & \ddots & \ddots & \ddots & \ddots & \ddots & \vdots \\ \vdots & & & \ddots & \ddots & \ddots & \ddots & \boldsymbol{O}_k \\ \vdots & & & & \ddots & \ddots & \ddots & \boldsymbol{B} \\ \boldsymbol{O}_k & \cdots & \cdots & \cdots & \cdots & \boldsymbol{O}_k & \boldsymbol{B}' & \boldsymbol{C} \end{bmatrix}, \qquad (12.12)$$

where $\boldsymbol{A}, \boldsymbol{B}$ and $\boldsymbol{C}$ are $k \times k$ matrices defined in (12.11).

13

Hierarchical Multivariate Time Series

13.1 Introduction

We describe how `R-INLA` can be used to fit hierarchical dynamic models for multivariate time series. Section 13.2 extends the Gaussian hierarchical models in Chapter 6 to multivariate (panel) time series, using ideas from Chapter 12. Section 13.3 describes a *level correlated model* (LCM) for fitting subject-specific models to multiple time series of counts.

Custom functions must be sourced by the user.

```
source("functions_custom.R")
```

We use the following packages.

```
library(mvtnorm)
library(INLA)
library(tidyverse)
```

We analyze the following data.

1. Ridesourcing in NYC

13.2 Multivariate hierarchical dynamic linear model

Response and predictor variables

For $t = 1, \ldots, n$ and $i = 1, \ldots, g$, let $\boldsymbol{y}_{it} = (Y_{1,it}, \ldots, Y_{q,it})'$ denote a q-dimensional response vector. We can view these g time series as a panel of vector-valued time series, so that for each i, we have a time series of q-dimensional vectors, each of length n. The response can be written as a data matrix $\boldsymbol{Y}$ with g rows (subjects/locations) and n columns (time points), where each element of the matrix is a q-dimensional vector:

$$\boldsymbol{Y} = \begin{pmatrix} \boldsymbol{y}_{11} & \boldsymbol{y}_{12} & \cdots & \boldsymbol{y}_{1n} \\ \boldsymbol{y}_{21} & \boldsymbol{y}_{22} & \cdots & \boldsymbol{y}_{2n} \\ \vdots & \vdots & \ddots & \vdots \\ \boldsymbol{y}_{g1} & \boldsymbol{y}_{g2} & \cdots & \boldsymbol{y}_{gn} \end{pmatrix}.$$

Let $j = 1, \ldots, q$ index the components of $\boldsymbol{y}_{it}$. We describe a general framework that involves different types of predictors and corresponding coefficients (similar to Chapter 6).

1. Predictors and coefficients that vary both with time t and location i: let $\boldsymbol{b}_{j,it} = (b_{j,it,1}, \ldots, b_{j,it,c})'$ be a c-dimensional vector of predictors, and let $\boldsymbol{\beta}_{j,it} = (\beta_{j,it,1}, \ldots, \beta_{j,it,c})'$ be the corresponding location and time-varying coefficient vector.

2. Predictors and coefficients that only vary with time t and not over locations i: let $\boldsymbol{d}_{j,t} = (d_{j,t,1}, \ldots, d_{j,t,a})'$ be an a-dimensional vector and let $\boldsymbol{\gamma}_{j,t} = (\gamma_{j,t,1}, \ldots, \gamma_{j,t,a})'$ be the corresponding time-varying (or dynamic) coefficient vector.

3. Predictors and coefficients that only vary with location i and not over time t: suppose that $\boldsymbol{s}_{j,i} = (s_{j,i,1}, \ldots, s_{j,i,b})'$ is a b-dimensional vector and $\boldsymbol{\eta}_{j,i} = (\eta_{j,i,1}, \ldots, \eta_{j,i,b})'$ is the corresponding location specific (or static) coefficient vector.

Note that we can generalize the above setup by allowing a, b, and c to be a_j, b_j, and c_j respectively for $j = 1, \ldots, q$, if the data complexity requires this (i.e., we include a different set of predictors for different components of the response vector).

Model setup

A hierarchical dynamic multivariate model for $\boldsymbol{y}_{it}$ which includes dynamic and static coefficients is shown componentwise for $j = 1, \ldots, q$ as

$$Y_{j,it} = \alpha_{j,i} + \boldsymbol{b}'_{j,it}\boldsymbol{\beta}_{j,it} + \boldsymbol{d}'_{j,t}\boldsymbol{\gamma}_{j,t} + \boldsymbol{s}'_{j,i}\boldsymbol{\eta}_{j,i} + v_{j,it}, \text{ where,} \tag{13.1}$$

$$\beta_{j,it,h} = \phi_j^{(\beta_h)}\beta_{j,i(t-1),h} + w_{j,it,h}^{(\beta)}, \ h = 1, \ldots, c, \tag{13.2}$$

$$\gamma_{j,t,\ell} = \phi_j^{(\gamma_\ell)}\gamma_{j,t-1,\ell} + w_{j,t,\ell}^{(\gamma_\ell)}, \ \ell = 1, \ldots, a. \tag{13.3}$$

In what follows, $i = 1, \ldots, g$ and $t = 1, \ldots, n$. For identifiability, we assume that if an intercept is allowed in the dynamic part, there is no intercept in the static portion, i.e., either $d_{j,it,1} = 1$, or $s_{j,i,1} = 1$. In order to incorporate the *dependence* between the q components of $\boldsymbol{y}_{it}$, we can assume that

(a) the observation error vector $\boldsymbol{v}_{it}$ follows a q-variate normal distribution with mean $\boldsymbol{0}$ and variance $\boldsymbol{V}$, for each i and t,

(b) for $h = 1, \ldots, c$, $\boldsymbol{w}_{it,h}^{(\beta)}$ follows a q-variate normal distribution with mean $\boldsymbol{0}$ and variance $\boldsymbol{W}^{(\beta_h)}$, for each i and t, and

(c) for $\ell = 1, \ldots, a$, $\boldsymbol{w}_{t,\ell}^{(\gamma)}$ follows a q-variate normal distribution with mean $\boldsymbol{0}$ and variance $\boldsymbol{W}^{(\delta_\ell)}$ for each t.

Note that $\boldsymbol{V}$, $\boldsymbol{W}^{(\beta_h)}$, $h = 1, \ldots, c$, and $\boldsymbol{W}^{(\gamma_\ell)}$, $\ell = 1, \ldots, a$ are positive definite (p.d.) covariance matrices. The prior specification for the covariance matrix $\boldsymbol{V}$ was discussed in Chapter 12, while the priors for other scalar coefficients were discussed in earlier chapters. We assume an inverse-Wishart prior for $\boldsymbol{V}$, i.e., the precision matrix $\boldsymbol{V}^{-1}$ follows a Wishart distribution. The *iid2d*, *iid3d*, *iid4d*, and *iid5d* options enable us to handle these in R-INLA when $q = 2, 3, 4$, or 5. State error covariances

$$\boldsymbol{W}^{(\beta_h)} = \text{diag}((\sigma_{11}^{(\beta_h)})^2, \ldots, (\sigma_{qq}^{(\beta_h)})^2), \ h = 1, \ldots, c, \text{and}$$

$$\boldsymbol{W}^{(\gamma_\ell)} = \text{diag}((\sigma_{11}^{(\gamma_\ell)})^2, \ldots, (\sigma_{qq}^{(\gamma_\ell)})^2), \ \ell = 1, \ldots, a$$

are assumed to be diagonal, and we can employ log-gamma priors for the marginal precisions, as discussed in Chapter 3.

13.2.1   Example: Analysis of TNC and Taxi as responses

We build hierarchical Gaussian DLMs similar to Chapter 6 to NYC ridesourcing data, where TNC and Taxi are components of a *bivariate* response vector $\boldsymbol{y}_{it} = (Y_{1,it}, Y_{2,it})'$ for zone i in week t. We again use scaled data for $n = 129$ weeks, but only consider $g = 36$ taxi zones in Manhattan. Figure 13.1 shows time series plots of the scaled (dividing each value by $10,000$) TNC and Taxi usage from nine zones in Manhattan.

```
ride <-
  read.csv("tnc_weekly_data.csv",
           header = TRUE,
           stringsAsFactors = FALSE)
k <- 10000
ride$tnc <- ride$tnc/k
ride$taxi <- ride$taxi/k
ride$subway <- ride$subway/k
# Filter by borough = Manhattan
ride.m <- ride %>%
  filter(borough == "Manhattan")
n <- n_distinct(ride.m$date)
g <- n_distinct(ride.m$zoneid)
```

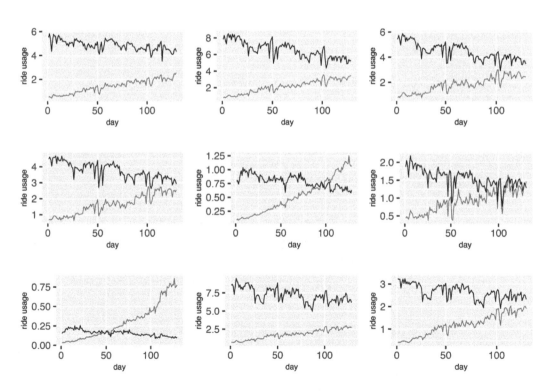

FIGURE 13.1: Weekly scaled TNC (red line) and Taxi (black line) usage in nine different taxi zones in Manhattan.

We reserve data from the last 5 weeks for each zone as hold-out data.

```
n.hold <- 5
n.train <- n - n.hold
df.zone <- ride.m %>%
  group_split(zoneid, )
test.df <- df.zone %>%
  lapply(function(x)
    tail(x, n.hold)) %>%
  bind_rows()
df.zone.append <- df.zone
for (i in 1:length(df.zone)) {
  df.zone.append[[i]]$tnc[(n.train + 1):n] <- NA
  df.zone.append[[i]]$taxi[(n.train + 1):n] <- NA
}
df.zone.append <- bind_rows(df.zone.append)
```

We stack the data by zone, while within each zone, we stack the data matrix corresponding to TNC usage followed by a data matrix for Taxi usage, as shown for the multivariate model setup in Chapter 12. We construct *subway.tnc* and *subway.taxi* in the following code.

```
y <- df.zone.append %>%
  group_split(zoneid) %>%
  lapply(function(x)
    unlist(select(x, tnc, taxi))) %>%
  unlist()
subway.tnc <- ride.m %>%
  select(zoneid, subway) %>%
  group_split(zoneid) %>%
  lapply(function(x)
    c(unlist(select(x, subway)), rep(NA, n))) %>%
  unlist()
subway.taxi <- ride.m %>%
  select(zoneid, subway) %>%
  group_split(zoneid) %>%
  lapply(function(x)
    c(rep(NA, n), unlist(select(x, subway)))) %>%
  unlist()
```

Although we do not show the code, we construct the predictors *holidays.tnc* and *holidays.taxi* by replacing "subway" by "holidays" in the code above. Similarly, we construct the predictors *precip.tnc* and *precip.taxi* by replacing "subway" by "precip".

The model is shown below; here $Y_{1,it}$ and $Y_{2,it}$ respectively denote TNC usage and Taxi usage.

$$Y_{1,it} = \alpha_{1,i} + \beta_{1,it,0} + \beta_{1,1}Z_{it,1} + \beta_{1,2}Z_{it,2} + \beta_{1,3}Z_{it,3} + v_{1,it},$$
$$Y_{2,it} = \alpha_{2,i} + \beta_{2,it,0} + \beta_{2,1}Z_{it,1} + \beta_{2,2}Z_{it,2} + \beta_{2,3}Z_{it,3} + v_{2,it}, \qquad (13.4)$$

with the observation error vector $\boldsymbol{v}_{it} \sim N(\boldsymbol{0}, \boldsymbol{V})$, $\boldsymbol{V}$ being a 2×2 p.d. covariance matrix. The state equation corresponding to the latent vector $\boldsymbol{\beta}_{it,0} = (\beta_{1,it,0}, \beta_{2,it,0})'$ is

$$\boldsymbol{\beta}_{it,0} = \boldsymbol{\Phi}\boldsymbol{\beta}_{i,t-1,0} + \boldsymbol{w}_{it}, \qquad (13.5)$$

where $w_{it} \sim N(0, W)$, and W and Φ are assumed to be diagonal. This corresponds to $q = 2$ independent latent $AR(1)$ processes for each zone $i = 1, \ldots, 36$, while Z_1, Z_2, and Z_3 correspond to the predictors Subway usage, holidays, and precipitation. Although we do not do this here, one may also include other predictors such as *scaled.population*, *scaled.employment*, *median.age*, and *median.earnings* that were discussed in Chapter 6.

Below, we discuss indexes and replicates that are common to Models L1–L3.

```
N <- 2 * n
t.index <- rep(c(1:N), g)
b0.tnc <- rep(c(1:n, rep(NA, n)), g)
b0.taxi <- rep(c(rep(NA, n), 1:n), g)
re.b0.taxi <- re.b0.tnc <- re.tindex <- rep(1:g, each = N)
```

Model L1: Same levels for TNC and Taxi usage across all zones

In (13.4), let $\alpha_{1,i} = \alpha_1$ and $\alpha_{2,i} = \alpha_2$ for $i = 1, \ldots, g$. The code below sets up the level indexes for this model.

```
alpha.tnc <- rep(c(rep(1, n), rep(NA, n)), g)
alpha.taxi <- rep(c(rep(NA, n), rep(1, n)), g)
```

The data is formatted as follows.

```
tnctaxi.hmv.data <- cbind.data.frame(
  y,
  t.index, re.tindex,
  alpha.tnc, alpha.taxi,
  b0.tnc, re.b0.tnc,
  b0.taxi, re.b0.taxi,
  subway.tnc, subway.taxi,
  holidays.tnc, holidays.taxi,
  precip.tnc, precip.taxi
)
```

The model formula is shown below, followed by the output.

```
formula.tnctaxi.hmv <-
  y  ~ -1 + alpha.tnc + alpha.taxi + subway.tnc + subway.taxi +
  holidays.tnc + holidays.taxi + precip.tnc + precip.taxi +
  f(t.index,
    model = "iid2d",
    n = N,
    replicate = re.tindex) +
  f(b0.tnc,
    model = "rw1",
    constr = FALSE,
    replicate = re.b0.tnc) +
  f(b0.taxi,
    model = "rw1",
    constr = FALSE,
    replicate = re.b0.taxi)
```

```
model.tnctaxi.hmv <-
  inla(
    formula.tnctaxi.hmv,
    family = c("gaussian"),
    data = tnctaxi.hmv.data,
    control.inla = list(h = 1e-5,
                        tolerance = 1e-3),
    control.compute = list(
      dic = TRUE,
      waic = TRUE,
      cpo = TRUE,
      config = TRUE
    ),
    control.predictor = list(compute = TRUE),
    control.family =
      list(hyper = list(prec = list(
        initial = 10,
        fixed = TRUE
      ))),
  )
```

Posterior summaries for the fixed and random effects are shown below. The observation error correlation R_{12} is significant, with a posterior mean of 0.776. The estimated α_1 and α_2 for TNC and Taxi usage are very close to zero. In the next section, we will investigate a model with different levels for TNC and Taxi usage for the different zones.

```
format.inla.out(model.tnctaxi.hmv$summary.fixed[,c(1,2,4)])
##   name          mean   sd      0.5q
## 1 alpha.tnc     0.001  31.626  0.000
## 2 alpha.taxi    0.001  31.626  0.000
## 3 subway.tnc    0.029  0.001   0.029
## 4 subway.taxi   0.064  0.001   0.064
## 5 holidays.tnc  0.024  0.006   0.024
## 6 holidays.taxi 0.028  0.013   0.028
## 7 precip.tnc    0.035  0.003   0.035
## 8 precip.taxi   0.008  0.006   0.008
format.inla.out(model.tnctaxi.hmv$summary.hyperpar[,c(1,2,4,5)])
##   name                                    mean     sd      0.5q     0.975q
## 1 Precision for t.index component 1       57.975   1.821   57.940   61.654
## 2 Precision for t.index component 2       13.565   0.391   13.557   14.355
## 3 Rho1:2 for t.index                      0.776    0.009   0.776    0.793
## 4 Precision for b0.tnc                    178.985  10.500  178.601  200.638
## 5 Precision for b0.taxi                   56.613   3.372   56.492   63.553
```

Fitted values and forecasts

Similar to Approach 1 in Chapter 12, we obtain fitted values and forecasts through posterior samples of the model parameters and hyperparameters using `inla.posterior.sample()`. We create a custom function `fit.post.sample.hmv()`.

```
fit.post.sample.hmv <- function(x) {
  set.seed(123457)
```

```
sigma2.v1 <- 1 / x$hyperpar[1]
sigma2.v2 <- 1 / x$hyperpar[2]
rho.v1v2 <- x$hyperpar[3]
cov.v1v2 <- rho.v1v2 * sqrt(sigma2.v1) * sqrt(sigma2.v2)
sigma.v <-
  matrix(c(sigma2.v1, cov.v1v2, cov.v1v2, sigma2.v2), nrow = 2)
b0.tnc <-
  matrix(x$latent[grep("b0.tnc", rownames(x$latent))], nrow = n)
b0.taxi <-
  matrix(x$latent[grep("b0.taxi", rownames(x$latent))], nrow = n)
fit.tnc <- fit.taxi <- vector("list", n)
for (j in 1:g) {
  V <- rmvnorm(n, mean = c(0, 0), sigma = sigma.v)
  V1 <- V[, 1]
  V2 <- V[, 2]
  fit.tnc[[j]] <-
    x$latent[grep("alpha.tnc", rownames(x$latent))] +
    b0.tnc[, j] +
    x$latent[grep("subway.tnc", rownames(x$latent))] *
    df.zone[[j]]$subway +
    x$latent[grep("holidays.tnc", rownames(x$latent))] *
    df.zone[[j]]$holidays +
    x$latent[grep("precip.tnc", rownames(x$latent))] *
    df.zone[[j]]$precip +
    V1
  fit.taxi[[j]] <-
    x$latent[grep("alpha.taxi", rownames(x$latent))] +
    b0.taxi[, j] +
    x$latent[grep("subway.taxi", rownames(x$latent))] *
    df.zone[[j]]$subway +
    x$latent[grep("holidays.taxi", rownames(x$latent))] *
    df.zone[[j]]$holidays +
    x$latent[grep("precip.taxi", rownames(x$latent))] *
    df.zone[[j]]$precip +
    V2
}
fit.tnc <-
  bind_cols(fit.tnc,
            .name_repair = ~ vctrs::vec_as_names(names =
                                                paste0("g", 1:g)))
fit.taxi <-
  bind_cols(fit.taxi,
            .name_repair = ~ vctrs::vec_as_names(names =
                                                paste0("g", 1:g)))
return(list(fit.tnc = fit.tnc, fit.taxi = fit.taxi))
}
```

The fitted values for scaled TNC and Taxi usage for each zone are computed by calling this function. Here, we demonstrate the computation for one zone (*zoneid* = "ID100"). The plots of observed and fitted and forecast values are shown in Figure 13.2.

```
fits <- post.sample.hmv %>%
  mclapply(function(x)
    fit.post.sample.hmv(x))
zone <- match("ID100", unique(df.zone.append$zoneid))
tnc.fit <- fits %>%
  lapply(function(x)
    x$fit.tnc[, zone]) %>%
  bind_cols() %>%
  rowMeans()
taxi.fit <- fits %>%
  lapply(function(x)
    x$fit.taxi[, zone]) %>%
  bind_cols() %>%
  rowMeans()
```

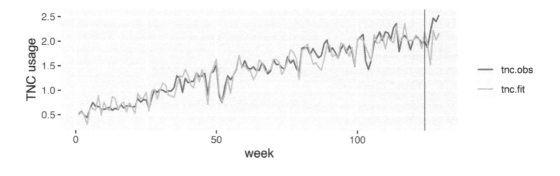

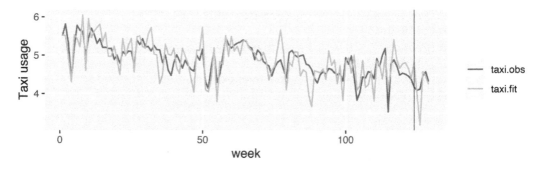

FIGURE 13.2: Observed (red) and fitted (blue) TNC (top plot) and Taxi usage for zone ID100 under Model L1. The black vertical line divides the data into training and test portions.

Model L2. Different fixed-effect zone-specific levels

We consider zone-specific levels $\alpha_{j,i}$, $j = 1, 2$ in (13.4). For a model with different *fixed* levels across zones, the indexes for $\alpha_{j,i}$ are defined below.

```
id.alpha.tnc <- id.alpha.taxi <- c()
for (i in 1:g) {
  temp.tnc <- c(rep(i, n), rep(NA, n))
```

```
  id.alpha.tnc <- c(id.alpha.tnc, temp.tnc)
  temp.taxi <- c(rep(NA, n), rep(i, n))
  id.alpha.taxi <- c(id.alpha.taxi, temp.taxi)
}
fact.alpha.tnc <- as.factor(id.alpha.tnc)
fact.alpha.taxi <- as.factor(id.alpha.taxi)
```

To get the formatted data tnctaxi.hmvfe.data for Model L1, we replace alpha.tnc and alpha.taxi by fact.alpha.tnc and fact.alpha.taxi in tnctaxi.hmv.data. The model formula formula.tnctaxi.hmvfe is the same as formula.tnctaxi.hmv except for replacing

```
y  ~ -1 + alpha.tnc + alpha.taxi + subway.tnc +...
```

by

```
y  ~ -1 + fact.alpha.tnc + fact.alpha.taxi + subway.tnc+...
```

We change the code

```
model.tnctaxi.hmv <-
  inla(
    formula.tnctaxi.hmv,
    family = c("gaussian"),
    data = tnctaxi.hmv.data, ...
```

in Model L1 to

```
model.tnctaxi.hmvfe <-
  inla(
    formula.tnctaxi.hmvfe,
    family = c("gaussian"),
    data = tnctaxi.hmvfe.data, ...
```

and obtain the following results.

```
##   name            mean  sd    0.5q
## 1 subway.tnc     0.029 0.001 0.029
## 2 subway.taxi    0.064 0.001 0.064
## 3 holidays.tnc   0.024 0.006 0.024
## 4 holidays.taxi  0.028 0.013 0.028
## 5 precip.tnc     0.035 0.003 0.035
## 6 precip.taxi    0.008 0.006 0.008
##   name                                 mean    sd     0.025q   0.975q
## 1 Precision for t.index component 1    58.030  1.817  54.531   61.681
## 2 Precision for t.index component 2    13.573  0.397  12.807   14.371
## 3 Rho1:2 for t.index                   0.775   0.009  0.757    0.793
## 4 Precision for b0.tnc                 178.705 10.325 159.478 199.958
## 5 Precision for b0.taxi                56.660  3.435  50.271   63.746
```

The posterior means for all the $\alpha_{j,i}$ are not significantly different than zero. The estimates of the other fixed effects and random effects are very similar to estimates from Model L1. The estimated correlation R_{12} is about 0.775. Under Model L1, we defined the custom function fit.post.sample.hmv() which uses inla.posterior.sample(). Here, we can use the same function to compute the fitted values with some minor changes, details of which are available on our GitHub link.

Model L3. Different random-effect zone-specific levels

We assume that $\alpha_{j,i} \sim N(0, \sigma_j^2)$ for $i = 1, \ldots, g$. The indexes for the random $\alpha_{j,i}$ are computed as follows.

```
id.alpha.tnc <- id.alpha.taxi <- c()
for (i in 1:g) {
  temp.tnc <- c(rep(i,n), rep(NA, n))
  id.alpha.tnc <- c(id.alpha.tnc, temp.tnc)
  temp.taxi <- c(rep(NA, n), rep(i, n))
  id.alpha.taxi <- c(id.alpha.taxi, temp.taxi)
}
```

The formatted data `tnctaxi.hmvre.data` is constructed by replacing `alpha.tnc` and `alpha.taxi` in `tnctaxi.hmv.data` under Model L1 by `id.alpha.tnc` and `id.alpha.taxi`.

The model formula is given below.

```
prec.prior <- list(prec = list(param = c(0.001, 0.001)))
formula.tnctaxi.hmvre <-
  y   ~ -1 +
  f(id.alpha.tnc, model = "iid", hyper = prec.prior) +
  f(id.alpha.taxi, model = "iid", hyper = prec.prior) +
  subway.tnc + subway.taxi + holidays.tnc + holidays.taxi +
  precip.tnc + precip.taxi +
  f(t.index,
    model = "iid2d",
    n = N,
    replicate = re.tindex) +
  f(b0.tnc,
    model = "rw1",
    constr = FALSE,
    replicate = re.b0.tnc) +
  f(b0.taxi,
    model = "rw1",
    constr = FALSE,
    replicate = re.b0.taxi)
```

We obtain results by changing the code for Model L1 from

```
model.tnctaxi.hmv <-
  inla(
    formula.tnctaxi.hmv,
    family = c("gaussian"),
    data = tnctaxi.hmv.data, ...
```

to the code below:

```
model.tnctaxi.hmvre <-
  inla(
    formula.tnctaxi.hmvre,
    family = c("gaussian"),
    data = tnctaxi.hmvre.data, ...
```

TABLE 13.1: Comparison of Models L1–L3.

Model	DIC	WAIC	Average Out-of-Sample MAE (TNC)	Average Out-of-Sample MAE(Taxi)
L1	−55035.97	−57758.129	0.222	0.336
L2	−55035.969	−57758.148	0.178	0.223
L3	−55035.963	−57758.201	0.176	0.228

```
##   name         mean  sd    0.5q
## 1 subway.tnc   0.029 0.001 0.029
## 2 subway.taxi  0.064 0.001 0.064
## 3 holidays.tnc 0.024 0.006 0.024
## 4 holidays.taxi 0.028 0.013 0.028
## 5 precip.tnc   0.035 0.003 0.035
## 6 precip.taxi  0.008 0.006 0.008
##   name                             mean    sd     0.5q    0.975q
## 1 Precision for id.alpha.tnc       63.743  44.756 51.855  184.685
## 2 Precision for id.alpha.taxi      79.883  73.026 58.269  277.321
## 3 Precision for t.index component 1 58.014 1.836  57.986  61.696
## 4 Precision for t.index component 2 13.571 0.396  13.565  14.364
## 5 Rho1:2 for t.index               0.775   0.010  0.776   0.794
## 6 Precision for b0.tnc             178.962 10.523 178.623 200.622
## 7 Precision for b0.taxi            56.701  3.440  56.596  63.766
```

The fitted values are computed using `inla.posterior.sample()`and the function `fit.sample.hmv.re()`. For details, see the GitHub link for the book.

In-sample and out-of-sample model comparisons for the three models are shown in Table 13.1. The DIC, WAIC, and MAE values (averaged over all zones) are similar across the three models. The simplest model is Model L1.

13.3 Level correlated models for multivariate time series of counts

In certain situations, we need to build a model for a multivariate time series of counts as a function of its history and available exogenous predictors. The dependence between the components may be due to omitted variables which simultaneously affect the response vector (Aitchison and Ho (1989); Chib and Winkelmann (2001)). Univariate count (Poisson or negative binomial) models for each component cannot account for the dependence among the components. See Soyer and Zhang (2021) for a detailed review of general ideas in Bayesian modeling of multivariate time series of counts.

Serhiyenko et al. (2018) described *level correlated models* (LCMs) which allow us to combine different component count distributions, accommodate their dependence, and offer a flexible modeling approach using `R-INLA`; also see Serhiyenko (2015) and Serhiyenko et al. (2015). Serhiyenko et al. (2016) and Wang et al. (2017) discussed crash counts modeling in transportation safety. Raman et al. (2021a) presented a application in marketing to assess effectiveness of promotions on product sales, while Wang et al. (2021) described an application for high-frequency financial data.

Let $\boldsymbol{y}_{it} = (y_{1,it}, \ldots, y_{q,it})'$ be a q-variate vector of count responses over time, for $i = 1, \ldots, g$ and $t = 1, \ldots, n$. That is, we observe q components of counts on g subjects (or at g locations) over n regularly spaced time points. As shown in Serhiyenko et al. (2018) or Raman et al. (2021a), we can write the LCM for the multivariate count time series as a state space model. In the following, $i = 1, \ldots, g$, $t = 1, \ldots, n$, and $j = 1, \ldots, q$.

13.3.1   Example: TNC and Taxi counts based on daily data

We look at daily TNC and Taxi counts over 3 years from 34 zones in Manhattan.

```
ride.m <-
  read.csv(
    "tnctaxi_daily_data_Manhattan.csv",
    header = TRUE,
    stringsAsFactors = FALSE
  )
n <- n_distinct(ride.m$date)
g1 <- n_distinct(ride.m$zoneid)
```

We hold out 14 days (two weeks) of data and set up the training and test data as follows.

```
n.hold <- 14 # 2 weeks
n.train <- n - n.hold
df.zone <- ride.m %>%
  group_split(zoneid)
test.df <- df.zone %>%
  lapply(function(x)
    tail(x, n.hold)) %>%
  bind_rows()
df.zone.append <- df.zone
for (i in 1:length(df.zone)) {
  df.zone.append[[i]]$tnc[(n.train + 1):n] <- NA
  df.zone.append[[i]]$taxi[(n.train + 1):n] <- NA
}
df.zone.append <- bind_rows(df.zone.append)
```

Model LCM1 for TNC and Taxi counts

Similar to Model Z1 in Chapter 9, we include zone-specific fixed levels $\alpha_{j,i}$ for each type j, a linear trend γt, and fixed effects corresponding to the predictors $\cos(2\pi t/7)$ and $\sin(2\pi t/7)$ (to handle the seasonality with period 7), and Subway usage. We denote these predictors by $Z_{t,\ell}$, $\ell = 1, \ldots, 3$. We include a dynamic intercept $\beta_{j,it,0}$ (this was denoted by $X_{j,it}$ in the general specification above) which is modeled as a random walk. We assume that the level correlated effect $\boldsymbol{\xi} = (\xi_{1,it}, \xi_{2,it})' \sim N(\boldsymbol{0}, \boldsymbol{\Sigma})$. The model is

$$Y_{j,it} | \lambda_{j,it} \sim \text{UDC}_j(\lambda_{j,it}),$$

$$\log \lambda_{j,it} = \alpha_{j,i} + \gamma t + \beta_{j,it,0} + \sum_{\ell=1}^{3} \beta_\ell Z_{t,\ell} + \xi_{j,it},$$

$$\beta_{j,it,0} = \beta_{j,i,t-1,0} + w_{j,it}, \tag{13.6}$$

where $w_{j,it}$ is $N(0, \sigma_j^2)$. The indexes and replicates are constructed in the code below.

```
N <- 2 * n
trend <- rep(1:n, 2)
htrend <- rep(trend, g1)
seas.per <- 7
costerm <- cos(2 * pi * trend / seas.per)
sinterm <- sin(2 * pi * trend / seas.per)
hcosterm <- rep(costerm, g1)
hsinterm <- rep(sinterm, g1)
#indexes
t.index <- rep(c(1:N), g1)
b0.tnc <- rep(c(1:n, rep(NA, n)), g1)
b0.taxi <- rep(c(rep(NA, n), 1:n), g1)
## replicates
re.b0.taxi <- re.b0.tnc <- re.tindex <- rep(1:g1, each = N)
id.alpha.tnc <- rep(1:g1, each = N)
fact.alpha.tnc <- as.factor(id.alpha.tnc)
id.alpha.taxi <- rep(1:g1, each = N)
fact.alpha.taxi <- as.factor(id.alpha.taxi)
```

The following code shows how to set up the predictor Subway usage corresponding to the TNC and Taxi counts.

```
y <- df.zone.append %>%
  group_split(zoneid) %>%
  lapply(function(x)
    unlist(select(x, tnc, taxi))) %>%
  unlist()
subway.tnc <- ride.m %>%
  select(zoneid, subway) %>%
  group_split(zoneid) %>%
  lapply(function(x)
    c(unlist(select(x, subway)), rep(NA, n))) %>%
  unlist()
subway.taxi <- ride.m %>%
  select(zoneid, subway) %>%
  group_split(zoneid) %>%
  lapply(function(x)
    c(rep(NA, n), unlist(select(x, subway)))) %>%
  unlist()
```

We collect the variables into a data frame `data.ride.lcm1`.

```
data.ride.lcm1 <-cbind.data.frame(y, t.index, htrend, hcosterm, hsinterm,
                                  fact.alpha.tnc, fact.alpha.taxi,
                                  b0.tnc, b0.taxi,
                                  re.b0.tnc, re.b0.taxi, re.tindex,
                                  subway.tnc, subway.taxi)
```

We assume Poisson observation models for both TNC and Taxi counts; i.e., we set UDC as Poisson. We set up the formula by assuming that the levels $\alpha_{j,i}$ are fixed effects, and use default prior specifications. The results are summarized below. Detailed output can be obtained from our GitHub link.

```
formula.ride.lcm1 <-
  y ~ -1 + fact.alpha.tnc + fact.alpha.taxi +
  htrend + hcosterm + hsinterm +
  subway.tnc + subway.taxi +
  f(t.index,
    model = "iid2d",
    n = N,
    replicate = re.tindex) +
  f(b0.tnc,
    model = "rw1",
    constr = FALSE,
    replicate = re.b0.tnc) +
  f(b0.taxi,
    model = "rw1",
    constr = FALSE,
    replicate = re.b0.taxi)

model.pois.lcm1 <- inla(formula.ride.lcm1,
      data = data.ride.lcm1,
      family="poisson",
      control.inla = list(h = 1e-5, tolerance = 1e-3),
      control.predictor = list(compute = TRUE, link = 1),
      control.compute = list(dic = TRUE, waic = TRUE,
                             cpo = TRUE, config = TRUE),
      control.fixed = list(expand.factor.strategy = "inla")
      )
summary.model.pois.lcm1 <- summary(model.pois.lcm1)

#format.inla.out(head(summary.model.pois.lcm1$fixed[,c(1,2, 5)]))
format.inla.out(summary.model.pois.lcm1$hyperpar[,c(1,2)])
##    name                                 mean      sd
## 1 Precision for t.index component 1     64.29     1.711
## 2 Precision for t.index component 2     66.01     1.165
## 3 Rho1:2 for t.index                     0.92     0.003
## 4 Precision for b0.tnc                3924.43   241.019
## 5 Precision for b0.taxi              10183.25   838.265
```

Posterior means for the levels $\alpha_{j,i}$ are very small ($< 1e - 05$). The details can be obtained by using `model.pois.lcm1$summary.fixed`.

We use `inla.posterior.sample()` to generate posterior samples of model parameters and hyperparameters, and construct distributions of fitted responses. Figure 13.3 shows plots of observed and fitted TNC and Taxi counts in zone ID107.

Since this setup involves likelihoods of two components, we can alternatively set up the responses in a *matrix form* and modify the formula and `inla()` call. Specifically, we replace

```
y <- df.zone.append %>%
  group_split(zoneid) %>%
  lapply(function(x) unlist(select(x, tnc, taxi))) %>%
  unlist()
```

in the code above by

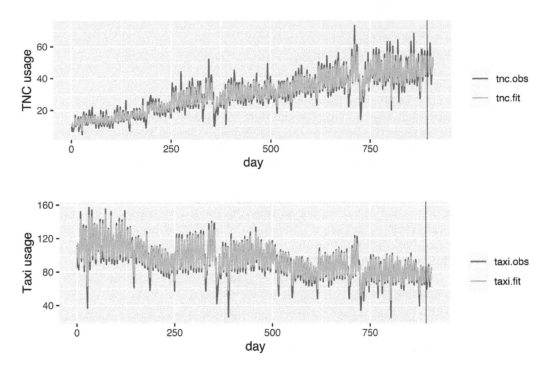

FIGURE 13.3: Observed (red) and fitted(blue) daily TNC and Taxi counts in taxi zone ID107 under Model LCM1. The black vertical line divides the data into training and test portions.

```
y <- df.zone.append %>%
  group_split(zoneid) %>%
  lapply(function(x) unlist(select(x, tnc, taxi))) %>%
  unlist()
y.alt.tnc <- df.zone.append %>%
  group_split(zoneid) %>%
  lapply(function(x) c(x$tnc, rep(NA, n))) %>%
  unlist()
y.alt.taxi <- df.zone.append %>%
  group_split(zoneid) %>%
  lapply(function(x) c(rep(NA, n), x$taxi)) %>%
  unlist()
Y <- matrix(c(y.alt.tnc, y.alt.taxi), ncol = 2)
```

We also replace y by Y in the data frame and formula. It can be verified that this code produces exactly the same results as the previous code. More details are given in our GitHub link.

```
model.pois.pois.lcm1 <- inla(
  formula.ride.lcm1,
  data = data.ride.lcm1,
  family = c("poisson", "poisson"),
```

```
   control.inla = list(h = 1e-5,
                     tolerance = 1e-3),
   control.predictor = list(compute = TRUE),
   control.compute = list(dic = TRUE, waic = TRUE, cpo = TRUE),
)
```

Based on the results, the Poisson distribution seems to be a good fit to the TNC and Taxi usage data. However, in other examples, count data may exhibit overdispersion and/or zero-inflation. R-INLA includes the negative binomial and zero-inflated Poisson as other options for observation models for counts. These can be invoked via `family = c("nbinomial")` or `family = c("zeroinflatedpoisson1")`. If all the components are assumed to follow the same UDC, then we can use the first code or the alternate code shown under Model LCM1. However, if the components follow different distributions, the alternate code (which uses the matrix Y instead of the vector y) must be used. For example, we can include `family = c("nbinomial", "poisson")` to assume that UDC is negative binomial for the first component and Poisson for the second component. Since neither the negative binomial nor the zero-inflated Poisson are suitable distributions for TNC or Taxi counts, we do not illustrate these here. See Serhiyenko et al. (2018) for the use of different UDC's with an application to marketing data.

Next, we consider a different model for TNC and Taxi counts instead of Model LCM1.

Model LCM2 for TNC and Taxi counts

We assume that the dynamic random effect or the levels do not vary by zone i. That is, $\alpha_{j,i} = \alpha_j$ and $\beta_{j,it,0} = \beta_{j,t,0}$ for all i.

$$Y_{j,it}|\lambda_{j,it} \sim \text{UDC}_j(\lambda_{j,it}),$$

$$\log \lambda_{j,it} = \alpha_j + \gamma_j t + \beta_{j,t,0} + \sum_{\ell=1}^{3} \beta_\ell Z_{t,\ell} + \xi_{j,it},$$

$$\beta_{j,t,0} = \beta_{j,t-1,0} + w_{j,t,0}, \tag{13.7}$$

where $w_{j,t}$ is $N(0, \sigma_w^2)$. Let UDC be the Poisson distribution. We again read in the daily data from 34 zones in Manhattan and set up the training and test sets. The indexes and replicates are shown below.

```
N <- 2 * n
trend <- rep(1:n, 2)
htrend <- rep(trend, g1)
seas.per <- 7
costerm <- cos(2 * pi * trend / seas.per)
sinterm <- sin(2 * pi * trend / seas.per)
hcosterm <- rep(costerm, g1)
hsinterm <- rep(sinterm, g1)
#indexes
t.index <- rep(c(1:N), g1)
b0.tnc <- id.re.b0.tnc <- rep(c(1:n, rep(NA, n)), g1)
b0.taxi <- id.re.b0.taxi <- rep(c(rep(NA, n), 1:n), g1)
## replicates
re.tindex <- rep(1:g1, each = N)
re.b0.taxi <- re.b0.tnc <- rep(rep(c(1, 2), each = n), each = g1)
```

```
id.alpha.tnc <- rep(1:g1, each = N)
fact.alpha.tnc <- as.factor(id.alpha.tnc)
id.alpha.taxi <- rep(1:g1, each = N)
fact.alpha.taxi <- as.factor(id.alpha.taxi)
```

We then set up the data frame.

```
y <- df.zone.append %>%
  group_split(zoneid) %>%
  lapply(function(x) unlist(select(x, tnc, taxi))) %>%
  unlist()
subway.tnc <- ride.m %>%
  select(zoneid, subway) %>%
  group_split(zoneid) %>%
  lapply(function(x) c(unlist(select(x,subway)), rep(NA, n))) %>%
  unlist()
subway.taxi <- ride.m %>%
  select(zoneid, subway) %>%
  group_split(zoneid) %>%
  lapply(function(x) c(rep(NA, n),unlist(select(x,subway)))) %>%
  unlist()
data.ride.lcm2 <-cbind.data.frame(y, t.index, htrend, hcosterm, hsinterm,
                                  id.alpha.tnc, id.alpha.taxi,
                                  b0.tnc, b0.taxi,
                                  re.b0.tnc, re.b0.taxi, re.tindex,
                                  subway.tnc, subway.taxi,
                                  id.re.b0.tnc, id.re.b0.taxi)
```

The formula and model output are shown below.

```
formula.ride.lcm2 <-
  y ~ -1 + fact.alpha.tnc + fact.alpha.taxi +
  htrend + hcosterm + hsinterm +
  subway.tnc + subway.taxi +
  f(t.index,
    model = "iid2d",
    n = N,
    replicate = re.tindex) +
  f(b0.tnc,
    model = "rw1",
    constr = FALSE) +
  f(id.re.b0.tnc, model = "iid",
    replicate = re.b0.tnc,
    hyper = list(prec=list(fixed=T, initial = 20))) +
  f(b0.taxi,
    model = "rw1",
    constr = FALSE) +
  f(id.re.b0.taxi, model="iid",
    replicate = re.b0.taxi,
    hyper = list(prec=list(fixed=T, initial = 20)))
```

```
##    name          mean    sd      0.975q
## 1 alpha.tnc      0.000  31.622  62.033
## 2 alpha.taxi    -0.001  31.618  62.024
## 3 htrend         0.000   0.001   0.001
## 4 hcosterm      -0.101   0.006  -0.089
## 5 hsinterm       0.099   0.006   0.110
## 6 subway.tnc     0.004   0.000   0.004

##    name                                    mean      sd
## 1 Precision for t.index component 1        3.554     0.028
## 2 Precision for t.index component 2        1.739     0.012
## 3 Rho1:2 for t.index                       0.949     0.001
## 4 Precision for b0.tnc                    235.549    34.807
## 5 Precision for b0.taxi                  2976.154   877.149
```

Plots of the observed and fitted responses are shown in Figure 13.4.

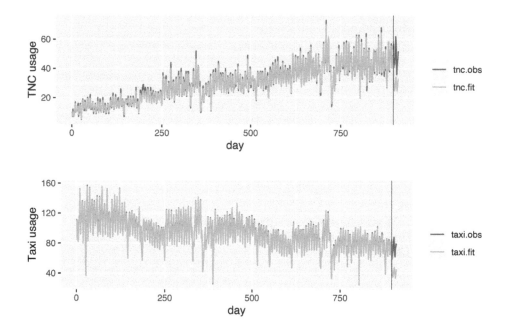

FIGURE 13.4: Observed (red) and fitted (blue) daily TNC and Taxi counts in taxi zone ID107 under Model LCM2.

Table 13.2 shows comparisons between Model LCM1 and Model LCM2. All criteria show that Model LCM1 performs much better than Model LCM2.

TABLE 13.2: Comparisons of DIC, WAIC, and MAE values from Models LCM1 and LCM2.

Model	DIC	WAIC	Average MAE (TNC)	Average MAE (Taxi)
LCM1	372026	365595	4.285	7.013
LCM2	419247	415004	13.815	30.997

14

Resources for the User

14.1 Introduction

Section 14.2 lists the R packages that we have used in the different chapters. Section 14.3 shows a few examples of custom functions we have developed to facilitate some repetitive calculations; details are available in the GitHub link associated with the book. A user must source these files before running the code in each chapter. Section 14.4 presents often used R-INLA items, pulled into one place for the user.

14.2 Packages used in the book

package	version	source
astsa	1.13	CRAN (R 4.1.0)
dlm	1.1-5	CRAN (R 4.1.0)
gridExtra	2.3	CRAN (R 4.1.0)
gtools	3.8.2	CRAN (R 4.1.0)
INLA	21.02.23	local
kableExtra	1.3.4	CRAN (R 4.1.0)
lubridate	1.7.10	CRAN (R 4.1.0)
mapview	2.9.0	CRAN (R 4.1.0)
marima	2.2	CRAN (R 4.1.0)
Matrix	1.3-3	CRAN (R 4.1.0)
matrixcalc	1.0-5	CRAN (R 4.1.0)
mvtnorm	1.1-1	CRAN (R 4.1.0)
quantmod	0.4.20	CRAN (R 4.1.2)
raster	3.4-10	CRAN (R 4.1.0)
readxl	1.3.1	CRAN (R 4.1.0)
sf	0.9-8	CRAN (R 4.1.0)
spdep	1.1-8	CRAN (R 4.1.0)
tables	0.9.6	CRAN (R 4.1.0)
tidyverse	1.3.1	CRAN (R 4.1.0)
tmap	3.3-1	CRAN (R 4.1.0)
vars	1.5-6	CRAN (R 4.1.0)

14.3   Custom functions used in the book

We show different custom functions that we have developed for the user. They are grouped
by functionality. The user must source the file containing these functions before running
the code, as we show in the beginning of each chapter in the book. All codes and datasets
used in the book can be accessed from https://github.com/ramanbala/dynamic-time-
series-models-R-INLA.

Basic plotting functions

```
multiline.plot <-
  function(plot.data,
           title = "",
           xlab = "",
           ylab = "",
           line.type = "solid",
           line.size = 1,
           line.color = "auto") {
    cpalette <-
      c("#000000",
        "#56B4E9",
        "#009E73",
        "#0072B2",
        "#D55E00",
        "#CC79A7")
    cname <- c("black", "blue", "green", "blue", "red", "purple")
    mp <- plot.data %>%
      pivot_longer(-time, names_to = "key", values_to = "dat")
    u <- unique(mp$key)
    mp$key <- factor(mp$key, levels = u)
    levels(mp$key) <- u
    mp.auto.color <- mp %>%
      ggplot(aes(x = time, y = dat, color = key)) +
      geom_line(size = line.size, linetype = line.type) +
      labs(x = xlab, y = ylab, colour = "") +
      ggtitle(title)
    mp.custom.color <- mp %>%
      ggplot(aes(x = time, y = dat, color = key)) +
      geom_line(size = line.size, linetype = line.type) +
      scale_color_manual(values = cpalette[match(line.color, cname)]) +
      labs(x = xlab, y = ylab, colour = "") +
      ggtitle(title)
  ifelse(line.color == "auto", return(mp.auto.color),
         return(mp.custom.color))
  }
```

Functions for forecast evaluation

```r
mae <- function(yhold,
                yfore,
                text = NA,
                round.digit = 3) {
  efore <- yhold - yfore
  mae <-  mean(abs(efore))
  if (is.na(text)) {
    return(paste("MAE is", round(mae, round.digit), sep = " "))
  } else {
    return(paste(text, round(mae, round.digit), sep = " "))
  }
}

mape <- function(yhold,
                 yfore,
                 text = NA,
                 round.digit = 3) {
  efore <- yhold - yfore
  mape <-  mean(abs(efore / yhold))
  if (is.na(text)) {
    return(paste("MAPE is", round(mape * 100, round.digit), sep = " "))
  } else {
    return(paste(text, round(mape * 100, round.digit), sep = " "))
  }
}
```

Function for model comparison

```r
# Model Selection Criteria - DIC, WAIC, PSBF and PIT
model.selection.criteria <- function(inla.result, plot.PIT = FALSE,
                                      n.train) {
  dic <- inla.result$dic$dic
  waic <- inla.result$waic$waic
  cpo <- inla.result$cpo$cpo
  psbf <- sum(log(cpo[1:n.train]))
  PIT <- inla.result$cpo$pit
  msc <- cbind(DIC=dic, WAIC = waic, PsBF = psbf)
  if(isTRUE(plot.PIT)){
    pit.hist <- hist(PIT, plot = F)
    return(list(msc = msc, hist=plot(pit.hist, main ="")))
  } else{
    return(msc = msc)
  }

}
```

Function for the filtering algorithm in DLM

```
filt.inla <-
  function(data.series,
           model = "rw1",
           ar.order = 0,
           trend = "no",
           alpha = "no") {
    pt <- proc.time()
    ### model formula
    eg <- expand.grid(
      model = c("rw1", "ar"),
      trend = c("yes", "no"),
      alpha = c("yes", "no")
    )
    eg <- apply(eg, 2, as.character)
    choice <- paste0(c(model, trend, alpha), collapse = ".")


    dat <- filt <- vector("list", length(data.series))
    for (k in 1:length(data.series)) {
      dat[[k]] <- data.series[1:k]
    }

    get.filter <- function(df, ar.order) {
      id <- trnd <- 1:length(df)
      inla.dat <- cbind.data.frame(data.series = df, trnd, id)
      ar.order <- as.numeric(ar.order)
      formula <- switch(
        choice,
        "rw1.no.no" = as.formula(
          paste0("data.series~-1+", "f(id, model='rw1', constr=FALSE)")
        ),
        "rw1.yes.no" = as.formula(
          paste0(
            "data.series~-1+trnd+",
            "f(id, model='rw1', constr=FALSE)"
          )
        ),
        "rw1.no.yes" = as.formula(
          paste0("data.series~ 1+", "f(id, model='rw1', constr=FALSE)")
        ),
        "rw1.yes.yes" = as.formula(
          paste0("data.series~ 1+trnd+", "f(id, model='rw1', constr=FALSE)")
        ),
        "ar.no.no" = as.formula(
          paste0("data.series~ -1+", "f(id, model='ar', order = ar.order)")
        ),
        "ar.yes.no" = as.formula(
          paste0(
            "data.series~ -1+trnd+",
```

```
              "f(id, model='ar', order = ar.order)"
          )
        ),
      "ar.no.yes" = as.formula(
          paste0("data.series~ 1+", "f(id, model='ar', order = ar.order)")
        ),
      "ar.yes.yes" = as.formula(
          paste0(
            "data.series~ 1+trnd+",
            "f(id, model='ar', order = ar.order)"
          )
        ),
    )
    model.inla <- inla(
      formula,
      family = "gaussian",
      data = inla.dat,
      control.predictor = list(compute = TRUE),
      control.compute = list(
        dic = T,
        config = TRUE,
        cpo = TRUE
      ),
      verbose = T
    )
    filt <- tail(model.inla$summary.random$id, 1)
    return(filt = filt)
}
#### model execution
require(doParallel)
require(foreach)
cores <-  detectCores()
if (cores > 2) {
  registerDoParallel(cores = detectCores() - 1)
}

if (ar.order > 1) {
  filt.all <-
    foreach (
      d = (2 + ar.order):length(data.series),
      .packages = c('tidyverse', 'INLA'),
      .verbose = T
    ) %dopar% {
      get.filter(dat[[d]], ar.order = ar.order)
    }
} else{
  filt.all <-
    foreach (
      d = 2:length(data.series),
      .packages = c('tidyverse', 'INLA'),
```

```
          .verbose = T
      ) %dopar% {
          get.filter(dat[[d]], ar.order = ar.order)
      }
  }
  filt.all.bind <- bind_rows(filt.all)
  # filt.est <- c(NA, filt.all$mean)
  inla.time <- proc.time() - pt
  return(list(filt.all.bind = filt.all.bind, time.taken = inla.time))
}
```

14.3.1   `rgeneric()` function for DLM-VAR model

```
inla.rgeneric.VAR_1 <-
  function(cmd = c("graph",
                   "Q",
                   "mu",
                   "initial",
                   "log.norm.const",
                   "log.prior",
                   "quit"),
           theta = NULL)
  {
    interpret.theta <- function() {
      n.phi <- k * k * p
      n.prec <- k
      n.tot <- n.phi + n.prec
      phi.VAR <-
        sapply(theta[as.integer(1:n.phi)], function(x) {
          x
        })
      # W matrix   precisions
      wprec <-
        sapply(theta[as.integer((n.phi + 1):(n.phi + n.prec))],
               function(x) {
                 exp(x)
               })
      param <- c(phi.VAR, wprec)
      W <- diag(1, n.prec)
      st.dev <- 1 / sqrt(wprec)
      st.dev.mat <-
        matrix(st.dev, ncol = 1) %*% matrix(st.dev, nrow = 1)
      W <- W * st.dev.mat
      PREC <- solve(W)
      return(list(
        param = param,
        VACOV = W,
        PREC = PREC,
        phi.vec = c(phi.VAR)
        ,
```

```r
    n.phi = n.phi,
    n.prec = n.prec
  ))
}
#Precision matrix
Q <- function() {
  param <- interpret.theta()
  phi.mat <- matrix(param$phi.vec, nrow = k)
  sigma.w.inv <- param$PREC
  A <- t(phi.mat) %*% sigma.w.inv %*% phi.mat
  B <- -t(phi.mat) %*% sigma.w.inv
  C <- sigma.w.inv
  # Construct mid-block:
  zero.mat <- matrix(0, nrow = 2, ncol = 2 * n)
  # Define the matrix block:
  mat <- cbind(t(B), A + C, B)
  # Initializing column id and matrix list:
  col.id <- list()
  mat.list <- list()
  col.id[[1]] <- 1:(3 * k)
  mat.list[[1]] <- zero.mat
  mat.list[[1]][, col.id[[1]]] <- mat
  for (id in 2:(n - 2)) {
    start.id <- col.id[[id - 1]][1] + k
    end.d <-  start.id + (3 * k - 1)
    col.id[[id]] <- start.id:end.d
    mat.list[[id]] <- zero.mat
    mat.list[[id]][, col.id[[id]]] <- mat
  }
  mid.block <- do.call(rbind, mat.list)
  tau.val <- 0.1
  diffuse.prec <- tau.val * diag(1, k)
  # Construct first and last row blocks and then join with mid block:
  first.row.block <-
    cbind(A + diffuse.prec, B, matrix(0, nrow = k,
  ncol = (k * n - k ^ 2)))
  last.row.block <-
    cbind(matrix(0, nrow = k, ncol = k * n - k ^ 2), t(B), C)
  toep.block.mat <-
    rbind(first.row.block, mid.block, last.row.block)
  # Changing to a sparse Matrix:
  prec.mat <- Matrix(toep.block.mat, sparse = TRUE)
  return(prec.mat)
}
# Graph function: Essentially Q matrix
graph = function() {
  return (inla.as.sparse(Q()))
}
#Mean of model
mu <- function() {
```

```r
    return(numeric(0))
}
# Log normal constant:
log.norm.const <- function() {
  Q <- Q()
  log.det.val <-
    Matrix::determinant(Q, logarithm = TRUE)$modulus
  val <- (-k * n / 2) * log(2 * pi) + 0.5 * log.det.val
  return (val)
}
log.prior <- function() {
  param <- interpret.theta()
  pars <- param$param
  k <- k
  total.par <- param$n.phi + param$n.prec
  # Normal prior for phi's:
  theta.phi <- theta[1:param$n.phi]
  phi.prior <-
    sum(sapply(theta.phi, function(x)
      dnorm(
        x,
        mean = 0,
        sd = 1,
        log = TRUE
      )))
  theta.prec <-
    theta[(param$n.phi + 1):(param$n.phi + param$n.prec)]
  prec.prior <-
    sum(sapply(theta.prec, function(x)
      dgamma(
        x,
        shape = 1,
        scale = 1,
        log = TRUE
      )))
  prec.jacob <- sum(theta.prec) # This is for precision terms
  prior.val <- phi.prior + prec.prior + prec.jacob
  return (prior.val)
}
initial = function() {
  phi.init <- c(0.1, 0.1, 0.1, 0.1)
  prec.init <- rep(1, k)
  init <- c(phi.init, prec.init)
  return (init)
}
if (as.integer(R.version$major) > 3) {
  if (!length(theta))
    theta <- initial()
} else {
  if (is.null(theta)) {
```

TABLE 14.2: Common control options – a quick lookup.

Options	Descriptions	Used in Chapter
control.compute	Specify what to actually compute during model fitting	Ch. 5-7
control.family	Specify model likelihood	Ch. 3-12
control.fixed	Options on fixed effects in a model	Ch. 3-4
control.group	Defines the order of the model, as in AR(p)	Ch. 3, 8
control.inla	Options on how the method is used in model fitting	Ch. 3, 12
control.lincomb	Options on how to compute linear combinations	Ch. 12
control.link		Ch. 12
control.predictor	Options on computing the linear predictor	Ch. 3-12

```
      theta <- initial()
    }
  }
  val <- do.call(match.arg(cmd), args = list())
  return (val)
}
```

14.4 Often used R-INLA items

Control options

Control options in the `inla()` function allow us to specify choices about the estimation process and output reporting. In Table 14.2, we list (in alphabetical order) a set of control options that are most relevant for dynamic time series modeling, along with their descriptions. Each control option has a manual page that can be accessed in the usual way, e.g., *?control.compute*, with detailed information about the option. In addition, the list of options for each control argument and their default values can be obtained with functions `inla.set.control.default()`, where CONTROL refers to the control argument. For example, for the argument *control.compute*, the list of options and default values can be obtained with `inla.set.control.compute.default()`. More details on these, as well as other `inla()` control options, can be seen on the R-INLA website.

Options for computing marginals

Often used options for computing various marginal distributions are shown in Table 14.3. We have used some of these options in various sections in the book.

Random effect specifications

R-INLA allows several specifications for handling random effects under different situations. Table 14.4 shows a collection of available effects from which a user may choose a suitable specification.

TABLE 14.3: Functions on marginals

Function	Description
inla.dmarginal	Compute the density function
inla.pmarginal	Compute the cumulative probability function
inla.qmarginal	Compute a quantile
inla.rmarginal	Draw a random sample
inla.hpdmarginal	Compute the highest posterior density (HPD) credible interval
inla.smarginal	Spline smoothing of the posterior marginal
inla.emarginal	Compute the expected value of a function
inla.mmarginal	Compute the posterior mode
inla.tmarginal	Transform a marginal using a function
inla.zmarginal	Compute summary statistics

TABLE 14.4: Random Effects in R-INLA.

Name	Description
iid	Independent and identically distributed Gaussian random effect
z	Classical specification of random effects
generic0	Generic specification (type 0)
generic1	Generic specification (type 1)
generic2	Generic specification (type 2)
generic3	Generic specification (type 3)
rw1	Random walk of order 1
rw2	Random walk of order 2
crw2	Continuous random walk of order 2
seasonal	Seasonal variation with a given periodicity
ar1	Autoregressive model of order 1
ar	Autoregressive model of arbitrary order
iid1d	Correlated random effects of dimension 1
iid2d	Correlated random effects of dimension
iid3d	Correlated random effects of dimension 3
iid4d	Correlated random effects of dimension 4
iid5d	Correlated random effects of dimension 5

Prior specifications

Table 14.5 shows a list of priors available in `R-INLA`. This table is similar to Table 5.2 in Gómez-Rubio (2020). These specifications enable a user to set priors for latent effects or hyperparameters. More details are available on the `R-INLA` website.

TABLE 14.5: Summary of priors implemented in R-INLA.

Prior	Description	Parameters
normal	Gaussian prior	mean, precision
gaussian	Gaussian prior	mean, precision
loggamma	log-Gamma prior	shape, rate
logtnormal	Truncated (positive) Gaussian prior	mean, precision
logtgaussian	Truncated (positive) Gaussian prior	mean, precision
flat	Flat (improper) prior on θ	
logflat	Flat (improper prior) on $\exp(\theta)$	
logiflat	Flat (improper prior) on $\exp(-\theta)$	
mvnorm	Multivarite Normal prior	
dirichlet	Dirichlet prior	α
betacorrelation	Beta prior for correlation	a, b
logitbeta	Beta prior, logit-scale	a, b
jeffreystdf	Jeffreys prior	
table	User defined prior	
expression	User defined prior	

Bibliography

Aitchison, J. and Ho, C. H. (1989). The multivariate Poisson-log normal distribution. *Biometrika*, 76(4):643–653.

Aktekin, T., Polson, N. G., and Soyer, R. (2020). A family of multivariate non-Gaussian time series models. *Journal of Time Series Analysis*, 41(5):691–721.

Andersson, J. (2001). On the normal inverse Gaussian stochastic volatility model. *Journal of Business & Economic Statistics*, 19(1):44–54.

Baker, S. G. (1994). The multinomial-Poisson transformation. *Journal of the Royal Statistical Society: Series D (The Statistician)*, 43(4):495–504.

Bakka, H., Rue, H., Fuglstad, G.-A., Riebler, A., Bolin, D., Illian, J., Krainski, E., Simpson, D., and Lindgren, F. (2018). Spatial modeling with R-INLA: a review. *Wiley Interdisciplinary Reviews: Computational Statistics*, 10(6):e1443.

Barndorff-Nielsen, O. E. (1997). Normal inverse Gaussian distributions and stochastic volatility modelling. *Scandinavian Journal of Statistics*, 24(1):1–13.

Barndorff-Nielsen, O. E. and Schou, G. (1973). On the parameterization of autoregressive models by partial autocorrelations. *Journal of Multivariate Analysis*, 3(4):408–419.

Berry, D. A. (1996). *Statistics: a Bayesian perspective*. Duxbury Press, Belmont, CA.

Berry, L. R. and West, M. (2020). Bayesian forecasting of many count-valued time series. *Journal of Business & Economic Statistics*, 38(4):872–887.

Besag, J. (1974). Spatial interaction and the statistical analysis of lattice systems. *Journal of the Royal Statistical Society: Series B (Methodological)*, 36(2):192–225.

Besag, J., York, J., and Mollié, A. (1991). Bayesian image restoration, with two applications in spatial statistics. *Annals of the Institute of Statistical Mathematics*, 43(1):1–20.

Blangiardo, M., Cameletti, M., Baio, G., and Rue, H. (2013). Spatial and spatio-temporal models with R-INLA. *Spatial and Spatio-Temporal Epidemiology*, 4:33–49.

Blei, D. M., Kucukelbir, A., and McAuliffe, J. D. (2017). Variational inference: a review for statisticians. *Journal of the American Statistical Association*, 112(518):859–877.

Boehmke, B. C. (2016). *Data Wrangling with R*. Springer International Publishing, Switzerland.

Bollerslev, T. (1986). Generalized autoregressive conditional heteroscedasticity. *Journal of Econometrics*, 31(3):307–327.

Cargnoni, C., Müller, P., and West, M. (1997). Bayesian forecasting of multinomial time series through conditionally gaussian dynamic models. *Journal of the American Statistical Association*, 92(438):640–647.

Chen, S. X. and Liu, J. S. (1997). Statistical applications of the Poisson-binomial and conditional Bernoulli distributions. *Statistica Sinica*, 7:875–892.

Chib, S. (1995). Marginal likelihood from the Gibbs output. *Journal of the American Statistical Association*, 90(432):1313–1321.

Chib, S. and Winkelmann, R. (2001). Markov chain Monte Carlo analysis of correlated count data. *Journal of Business & Economic Statistics*, 19(4):428–435.

Cox, D. R., Gudmundsson, G., Lindgren, G., Bondesson, L., Harsaae, E., Laake, P., Juselius, K., and Lauritzen, S. L. (1981). Statistical analysis of time series: some recent developments [with discussion and reply]. *Scandinavian Journal of Statistics*, 8:93–115.

Cressie, N. (2015). *Statistics for Spatial Data*. John Wiley & Sons, New York.

De Bruijn, N. G. (1981). *Asymptotic Methods in Analysis*. Dover Publications, New York.

Dutta, C., Ravishanker, N., and Basu, S. (2022). Modeling multivariate positive-valued time series using R-INLA. arXiv:2206.05374.

Ebrahimi, N., Soofi, E. S., and Soyer, R. (2010). On the sample information about parameter and prediction. *Statistical Science*, 25(3):348–367.

Engle, R. F. (1982). Autoregressive conditional heteroscedasticity with estimates of the variance of United Kingdom inflation. *Econometrica: Journal of the Econometric Society*, pages 987–1007.

Fahrmeir, L. and Tutz, G. (2001). Models for multicategorical responses: multivariate extensions of generalized linear models. In *Multivariate Statistical Modelling Based on Generalized Linear Models*, pages 69–137. Springer.

Fuentes, M., Chen, L., and Davis, J. M. (2008). A class of nonseparable and nonstationary spatial temporal covariance functions. *Environmetrics: The Official Journal of the International Environmetrics Society*, 19(5):487–507.

Fuglstad, G.-A., Simpson, D., Lindgren, F., and Rue, H. (2019). Constructing priors that penalize the complexity of Gaussian random fields. *Journal of the American Statistical Association*, 114(525):445–452.

Gamerman, D. (1998). Markov chain Monte Carlo for dynamic generalised linear models. *Biometrika*, 85(1):215–227.

Gamerman, D. and Lopes, F. H. (2006). *Markov Chain Monte Carlo: Stochastic Simulation for Bayesian Inference*. Chapman & Hall/CRC, New York.

Gamerman, D. and Migon, H. S. (1993). Dynamic hierarchical models. *Journal of the Royal Statistical Society: Series B (Methodological)*, 55(3):629–642.

Gelfand, A. E. and Dey, D. K. (1994). Bayesian model choice: asymptotics and exact calculations. *Journal of the Royal Statistical Society: Series B (Methodological)*, 56(3):501–514.

Gelfand, A. E. and Smith, A. F. M. (1990). Sampling-based approaches to calculating marginal densities. *Journal of the American Statistical Association*, 85:398–409.

Gelman, A. and Hill, J. (2006). *Data Analysis Using Regression and Multilevel/Hierarchical Models*. Cambridge University Press, New York.

Gelman, A., Hwang, J., and Vehtari, A. (2014). Understanding predictive information criteria for Bayesian models. *Statistics and Computing*, 24(6):997–1016.

Gerte, R., Konduri, K. C., Ravishanker, N., Mondal, A., and Eluru, N. (2019). Understanding the relationships between demand for shared ride modes: case study using open data from New York City. *Transportation Research Record*, 2673(12):30–39.

Gómez-Rubio, V. (2020). *Bayesian Inference with INLA*. CRC Press, Boca Raton, Florida.

Gómez-Rubio, V., Bivand, R. S., and Rue, H. (2020). Bayesian model averaging with the integrated nested Laplace approximation. *Econometrics*, 8(2):23.

Gómez-Rubio, V. and Rue, H. (2018). Markov chain Monte Carlo with the integrated nested Laplace approximation. *Statistics and Computing*, 28(5):1033–1051.

Guimaraes, P. (2004). Understanding the multinomial-Poisson transformation. *The Stata Journal*, 4(3):265–273.

Harvey, A. and Koopman, S. J. (2014). Structural time series models. *Wiley StatsRef: Statistics Reference Online*.

Held, L., Schrödle, B., and Rue, H. (2010). Posterior and cross-validatory predictive checks: a comparison of MCMC and INLA. In *Statistical Modelling and Regression Structures*, pages 91–110. Springer.

Hu, S., Ivan, J. N., Ravishanker, N., and Mooradian, J. (2013). Temporal modeling of highway crash counts for senior and non-senior drivers. *Accident Analysis & Prevention*, 50:1003–1013.

Hubin, A. and Storvik, G. (2016). Estimating the marginal likelihood with integrated nested Laplace approximation (INLA).

Hyndman, R., Athanasopoulos, G., Bergmeir, C., Caceres, G., Chhay, L., O'Hara-Wild, M., Petropoulos, F., Razbash, S., Wang, E., and Yasmeen, F. (2021). *forecast: Forecasting functions for time series and linear models*. R package version 8.15.

Hyndman, R. J. and Khandakar, Y. (2008). Automatic time series forecasting: the forecast package for R. *Journal of Statistical Software*, 26(3):1–22.

Jacquier, E., Polson, N. G., and Rossi, P. E. (1994). Bayesian analysis of stochastic volatility models. *Journal of Business & Economic Statistics*, pages 371–389.

Jeffreys, H. (1935). Some tests of significance, treated by the theory of probability. In *Mathematical Proceedings of the Cambridge Philosophical Society*, volume 31, pages 203–222. Cambridge University Press.

Jordan, M. I., Ghahramani, Z., Jaakkola, T. S., and Saul, L. K. (1998). An introduction to variational methods for graphical models. In *Learning in Graphical Models*, pages 105–161. Springer.

Kalman, R. E. (1960). A new approach to linear filtering and prediction problems. *Trans. ASME J. Basic Engineering*, 83.

Kass, R. E. and Raftery, A. E. (1995). Bayes factors. *Journal of the American Statistical Association*, 90(430):773–795.

Kim, S., Shephard, N., and Chib, S. (1998). Stochastic volatility: likelihood inference and comparison with ARCH models. *The Review of Economic Studies*, 65(3):361–393.

Knorr-Held, L. (2000). Bayesian modelling of inseparable space-time variation in disease risk. *Statistics in Medicine*, 19(17-18):2555–2567.

Korobilis, D. and Koop, G. (2018). Variational Bayes inference in high-dimensional time-varying parameter models.

Krainski, E. T., Gómez-Rubio, V., Bakka, H., Lenzi, A., Castro-Camilo, D., Simpson, D., Lindgren, F., and Rue, H. (2018). *Advanced Spatial Modeling with Stochastic Partial Differential Equations using R and INLA.* CRC Press, New York.

Kullback, S. (1959). *Information Theory and Statistics.* John Wiley & Sons, New York.

Kullback, S. and Leibler, R. A. (1951). On information and sufficiency. *The Annals of Mathematical Statistics*, 22(1):79–86.

Lauritzen, S. L. (1996). *Graphical Models*, volume Oxford Statistical Science Series 17. Clarendon Press, Oxford.

Lee, J. Y. L., Green, P. J., and Ryan, L. M. (2017). On the "Poisson trick" and its extensions for fitting multinomial regression models. *arXiv preprint arXiv:1707.08538.*

Lindgren, F. and Rue, H. (2015). Bayesian spatial modelling with R-INLA. *Journal of Statistical Software*, 63(19):1–25.

Lindley, D. V. (1961). The use of prior probability distributions in statistical inference and decision. In *Proc. 4th Berkeley Symp. on Math. Stat. and Prob*, pages 453–468.

Lindley, D. V. (1980). Approximate Bayesian methods. *Trabajos de estadística y de investigación operativa*, 31(1):223–245.

Lindley, D. V. (1983). Theory and practice of Bayesian statistics. *Journal of the Royal Statistical Society. Series D (The Statistician)*, 32(1/2):1–11.

Lunn, D. J., Thomas, A., Best, N., and Spiegelhalter, D. (2000). Winbugs-a bayesian modelling framework: concepts, structure, and extensibility. *Statistics and computing*, 10(4):325–337.

Lütkepohl, H. (2013). *Introduction to Multiple Time Series Analysis.* Springer Berlin, Heidelberg.

Marin, J.-M., Pudlo, P., Robert, C. P., and Ryder, R. J. (2012). Approximate Bayesian computational methods. *Statistics and Computing*, 22(6):1167–1180.

Marriott, J., Ravishanker, N., Gelfand, A. E., and Pai, J. S. (1996). Bayesian analysis of ARMA processes: complete sampling-based inference under exact likelihoods. *Bayesian Analysis in Statistics and Econometrics*, pages 243–256.

Martino, S., Aas, K., Lindqvist, O., Neef, L. R., and Rue, H. (2011). Estimating stochastic volatility models using integrated nested Laplace approximations. *The European Journal of Finance*, 17(7):487–503.

Martino, S. and Riebler, A. (2014). Integrated nested Laplace approximations (INLA). *Wiley StatsRef: Statistics Reference Online*, pages 1–19.

Martins, T. G., Simpson, D., Lindgren, F., and Rue, H. (2013). Bayesian computing with INLA: new features. *Computational Statistics & Data Analysis*, 67:68–83.

McCullagh, P. and Nelder, J. A. (1989). *Generalized Linear Models.* Chapman & Hall, London, 2nd edition.

Monahan, J. F. (1984). A note on enforcing stationarity in autoregressive-moving average models. *Biometrika*, 71(2):403–404.

Moraga, P. (2019). *Geospatial Health Data: Modeling and Visualization with R-INLA and Shiny*. CRC Press, Boca Raton, Florida.

Musa, J. D. (1979). Software reliability data. *IEEE Comput. Soc. Repository*.

Newton, M. A. and Raftery, A. E. (1994). Approximate Bayesian inference with the weighted likelihood bootstrap. *Journal of the Royal Statistical Society: Series B (Methodological)*, 56(1):3–26.

Palmí-Perales, F., Gómez-Rubio, V., and Martinez-Beneito, M. A. (2021). Bayesian multivariate spatial models for lattice data with INLA. *Journal of Statistical Software*, 98:1–29.

Petris, G. (2010). An R package for Dynamic Linear Models. *Journal of Statistical Software*, 36(12):1–16.

Plummer, M. et al. (2003). Jags: A program for analysis of bayesian graphical models using gibbs sampling. In *Proceedings of the 3rd international workshop on distributed statistical computing*, volume 124, pages 1–10. Vienna, Austria.

Raman, B., Ravishanker, N., Soyer, R., Gorti, V., and Sen, K. (2021a). Dynamic Bayesian modeling of count time series using R-INLA. *Journal of the Indian Statistical Association*, 58(2):157–192.

Raman, B., Sen, K., Gorti, V., and Ravishanker, N. (2021b). Improving promotional effectiveness for consumer goods—a dynamic bayesian approach. *Applied Stochastic Models in Business and Industry*, 37(4):823–833.

RStudio Team (2021). *RStudio: Integrated Development Environment for R*. RStudio, PBC, Boston, MA.

Rue, H. and Held, L. (2005). *Gaussian Markov Random Fields: Theory and Applications*. Chapman & Hall/CRC, Boca Raton, Florida.

Rue, H., Martino, S., and Chopin, N. (2009). Approximate Bayesian inference for latent Gaussian models using integrated nested Laplace approximations (with discussion). *Journal of the Royal Statistical Society, Series B*, 71:319–392.

Ruiz-Cárdenas, R., Krainski, E. T., and Rue, H. (2012). Direct fitting of dynamic models using integrated nested Laplace approximations—INLA. *Computational Statistics & Data Analysis*, 56(6):1808–1828.

Seppä, K., Rue, H., Hakulinen, T., Läärä, E., Sillanpää, M. J., and Pitkäniemi, J. (2019). Estimating multilevel regional variation in excess mortality of cancer patients using integrated nested Laplace approximation. *Statistics in Medicine*, 38(5):778–791.

Serhiyenko, V. (2015). Dynamic modeling of multivariate counts – fitting, diagnostics, and applications. *Ph.D. thesis, University of Connecticut*.

Serhiyenko, V., Mamun, S. A., Ivan, J. N., and Ravishanker, N. (2016). Fast Bayesian inference for modeling multivariate crash counts. *Analytic Methods in Accident Research*, 9:44–53.

Serhiyenko, V., Ravishanker, N., and Venkatesan, R. (2015). Approximate Bayesian estimation for multivariate count time series models. In *Ordered Data Analysis, Modeling and Health Research Methods*, pages 155–167. Springer.

Serhiyenko, V., Ravishanker, N., and Venkatesan, R. (2018). Multi-stage multivariate modeling of temporal patterns in prescription counts for competing drugs in a therapeutic category. *Applied Stochastic Models in Business and Industry*, 34(1):61–78.

Shumway, R. H. and Stoffer, D. S. (2017). *Time Series Analysis and Its Applications*. Springer Texts in Statistics. Springer International Publishing, New York.

Singpurwalla, N. D. and Soyer, R. (1992). Non-homogeneous autoregressive processes for tracking (software) reliability growth, and their Bayesian analysis. *Journal of the Royal Statistical Society: Series B (Methodological)*, 54(1):145–156.

Soyer, R., Aktekin, T., and Kim, B. (2016). Bayesian modeling of time series of counts with business applications. In *Handbook of Discrete-Valued Time Series*, pages 265–284. Chapman and Hall/CRC: New York.

Soyer, R. and Sung, M. (2013). Bayesian dynamic probit models for the analysis of longitudinal data. *Computational Statistics & Data Analysis*, 68:388–398.

Soyer, R. and Zhang, D. (2021). Bayesian modeling of multivariate time series of counts. *Wiley Interdisciplinary Reviews: Computational Statistics*, page e1559.

Spiegelhalter, D. J., Best, N. G., Carlin, B. P., and Van Der Linde, A. (2002). Bayesian measures of model complexity and fit. *Journal of the Royal Statistical Society: Series B (Statistical Methodology)*, 64(4):583–639.

Tanner, M. and Wong, W. H. (1987). The calculation of posterior distributions by data augmentation (with discussion). *Journal of the American Statistical Association*, 82:528–554.

Taylor, S. J. (1982). Financial returns modelled by the product of two stochastic processes – a study of the daily sugar prices 1961-75. *Time Series Analysis: Theory and Practice*, 1:203–226.

Team, R. C. (2018). *R: A Language and Environment for Statistical Computing*. R Foundation for Statistical Computing, Vienna, Austria.

Tierney, L. and Kadane, J. B. (1986). Accurate approximations for posterior moments and marginal densities. *Journal of the American Statistical Association*, 81(393):82–86.

Toman, P., Zhang, J., Ravishanker, N., and Konduri, K. C. (2020). Spatiotemporal analysis of ridesourcing and taxi demand by taxi zones in New York City. *Journal of the Indian Society for Probability and Statistics*, 22:231–249.

Tsay, R. S. (2005). *Analysis of Financial Time Series*. John Wiley & Sons, New York.

Van Niekerk, J., Bakka, H., Ruc, H., and Schenk, L. (2019). New frontiers in Bayesian modeling using the INLA package in R. *arXiv preprint arXiv:1907.10426*.

Wang, K., Ivan, J. N., Ravishanker, N., and Jackson, E. (2017). Multivariate Poisson lognormal modeling of crashes by type and severity on rural two lane highways. *Accident Analysis & Prevention*, 99:6–19.

Wang, X., Yue, Y., and Faraway, J. J. (2018). *Bayesian Regression Modeling with INLA*. Chapman and Hall/CRC, New York.

Wang, Y., Zou, J., and Ravishanker, N. (2021). Modeling correlated count time series using R-INLA. *Technical Report*.

Watanabe, S. and Opper, M. (2010). Asymptotic equivalence of Bayes cross validation and widely applicable information criterion in singular learning theory. *Journal of Machine Learning Research*, 11(12).

West, M. and Harrison, J. (1997). *Bayesian Forecasting and Dynamic Models*. Springer-Verlag, New York, 2 edition.

Wichern, D. W. and Jones, R. H. (1977). Assessing the impact of market disturbances using intervention analysis. *Management Science*, 24(3):329–337.

Wickham, H. (2016). *ggplot2: Elegant Graphics for Data Analysis*. Springer-Verlag, New York.

Wickham, H., Averick, M., Bryan, J., Chang, W., McGowan, L. D., François, R., Grolemund, G., Hayes, A., Henry, L., Hester, J., Kuhn, M., Pedersen, T. L., Miller, E., Bache, S. M., Müller, K., Ooms, J., Robinson, D., Seidel, D. P., Spinu, V., Takahashi, K., Vaughan, D., Wilke, C., Woo, K., and Yutani, H. (2019). Welcome to the tidyverse. *Journal of Open Source Software*, 4(43):1686.

Wikle, C. K., Berliner, M. L., and Cressie, N. (1998). Hierarchical Bayesian space-time models. *Environmental and Ecological Statistics*, 5(2):117–154.

Wood, S. N. (2020). Simplified integrated nested Laplace approximation. *Biometrika*, 107(1):223–230.

Xie, Y. (2016). *Bookdown: Authoring Books and Technical Documents with R Markdown*. CRC Press, Boca Raton, Florida.

Index